RDH · 33-1
TN269.D48

WESTERN GEOPHYSICAL

DEVELOPMENTS IN
GEOPHYSICAL EXPLORATION METHODS—1

THE DEVELOPMENTS SERIES

Developments in many fields of science and technology occur at such a pace that frequently there is a long delay before information about them becomes available and usually it is inconveniently scattered among several journals.

Developments Series books overcome these disadvantages by bringing together within one cover papers dealing with the latest trends and developments in a specific field of study and publishing them within *six months* of their being written.

Many subjects are covered by the series including food science and technology, polymer science, civil and public health engineering, pressure vessels, composite materials, concrete, building science, petroleum technology, geology, etc.

Information on other titles in the series will gladly be sent on application to the publisher.

DEVELOPMENTS IN GEOPHYSICAL EXPLORATION METHODS—1

Edited by

A. A. FITCH

Consultant, Formerly of Seismograph Service (England) Limited, Keston, Kent, UK

APPLIED SCIENCE PUBLISHERS LTD
LONDON

APPLIED SCIENCE PUBLISHERS LTD
RIPPLE ROAD, BARKING, ESSEX, ENGLAND

British Library Cataloguing in Publication Data

Developments in geophysical exploration methods.
1.
1. Prospecting—Geophysical methods
I. Fitch, A A
622'.15   TN269

ISBN 0-85334-835-9

WITH 7 TABLES AND 180 ILLUSTRATIONS

© APPLIED SCIENCE PUBLISHERS LTD 1979

All rights reserved. No part of this publication may be reproduced, stored in a retrieval system, or transmitted in any form or by any means, electronic, mechanical, photocopying, recording, or otherwise, without the prior written permission of the publishers, Applied Science Publishers Ltd, Ripple Road, Barking, Essex, England

Printed in Great Britain by Galliard (Printers) Ltd, Great Yarmouth

PREFACE

This is a collection of original papers, each by an expert in his field. They deal with different sectors of recent geophysical development. It may be, at first, difficult to see what else unites them, and how these several technologies can contribute to an integrated exploration process.

What brings these writers together is that they have all contributed to the improvement of what comes to the eye of the geophysical interpreter. Some of the improvement is achieved at the data-gathering stage, some of it in processing, and in presentation. For all of this improvement interpreters in general are most grateful. The editor is appreciative in a quite personal way, not only of the advances in technology, but also of the effort in writing which has been made by these busy contributors, and so created this collection.

Something can be said here about interpretation and the environment in which it is carried out, since it represents the field where the results of these technical developments are ultimately tested.

In the commercial world it is from the geophysical interpreter that management learns the results of a large sector of exploration expenditure, and learns them in a form on which still larger expenditures on later phases of exploration can be based.

The nature of the interpreter's work is inter-disciplinary. The techniques of geophysical interpretation are set uneasily among the classical divisions of science. Data gathering and processing employ the methods and language of the physical sciences and of mathematics. The earth itself is represented by generalised and simplified concepts, since the real earth is too complex and too varied for local or 'custom-built' treatment. These simplified models are the basis for instrument design, the design of field procedures and, above all, for seismic data processing. Even the simplified models which are used give rise to mathematical problems which are intractable, and a further stage of mathematical simplification is introduced into some subjects. Not always, in the literature, does one find a study of what simplification has done to the earth model, and whether this new model is more or less like the earth as we know it.

The interpreter, then, working with the output from these complex data gathering and processing procedures, has to derive a three-dimensional concept of the earth as it is, and to communicate it. He can rarely achieve the exact observation of a geologist working on surface, underground, or in a well; but he strives always to approach it.

Geophysical interpretation which has made full use of the known geology is the starting point for further chains of inference. In petroleum exploration the presence and attitude of source rocks, reservoir and cap rocks can be assessed. History of sedimentation, folding and faulting, and the formation and migration of hydrocarbons can be reconstructed. These are the considerations leading to well locations.

In coalmining the interest is in structure, especially small faults. The concern is not so much the discovery and valuation of coal, but in the assessment of how it should be mined.

In civil engineering the interest is rather to extend knowledge of structure and of strength of materials of a proposed structure downwards into the earth with which the structure is to be integrated. This is one of the least exploited of all the interpretation skills.

Scientific investigations are concerned with detailing the sub-surface geology; and the objective is achieved when the interpretation of the geophysical survey is securely integrated with the known geology.

Finally, my thanks to all of the writers who have contributed their thoughts and words for this volume.

A. A. FITCH
Limpsfield,
Surrey, UK

CONTENTS

Preface v

List of Contributors ix

1. Velocity Determination from Seismic Reflection Data . . 1
   M. AL-CHALABI

2. Patterns of Sources and Detectors 69
   S. D. BRASEL

3. Well Geophone Surveys and the Calibration of Acoustic Velocity Logs 93
   P. KENNETT

4. Seismic Sources on Land 115
   W. E. LERWILL

5. Marine Seismic Sources 143
   R. LUGG

6. Gravity and Magnetic Surveys at Sea 205
   L. L. NETTLETON

7. Pulse Shaping Methods 239
   D. G. STONE

8. Seismic Profiling for Coal on Land 271
   A. ZIOLKOWSKI

Index 307

LIST OF CONTRIBUTORS

M. AL-CHALABI

   Geophysicist, The British Petroleum Co. Ltd (EPD), Britannic House, Moor Lane, London EC2Y 9BU, UK.

S. D. BRASEL

   President and Consulting Geophysicist, Seismic International Research Corp., 818 17th Street, Suite 410, Denver, Colorado 80202, USA.

P. KENNETT

   Manager, Well Survey Division, Seismograph Service (England) Ltd, Holwood, Westerham Road, Keston, Kent BR2 6HD, UK.

W. E. LERWILL

   Senior Research Engineer, Seismograph Service (England) Ltd, Holwood, Westerham Road, Keston, Kent BR2 6HD, UK.

R. LUGG

   Marine Research and Development Geophysicist, Seismograph Service (England) Ltd, Holwood, Westerham Road, Keston, Kent BR2 6HD, UK.

L. L. NETTLETON

   Lecturer Emeritus in Geology, School of Natural Sciences, Department of Geology, Rice University, Houston, Texas 77001, USA.

D. G. STONE

Assistant Vice-President, Seismograph Service Corporation, PO Box 1590, Tulsa, Oklahoma 74102, USA.

A. ZIOLKOWSKI

Chief Geophysicist, Mining Department, National Coal Board, Hobart House, Grosvenor Place, London SW1X 7AE, UK.

Chapter 1

VELOCITY DETERMINATION FROM SEISMIC REFLECTION DATA

M. AL-CHALABI†

The British Petroleum Co. Limited, London, UK

SUMMARY

The determination of the gross velocity distribution in the ground from CDP‡ reflection data depends primarily on the determination of the velocity producing maximum coherency in the stacked data. This stacking velocity is a mathematical quantity which has no physical significance. The interval and average velocities describe meaningful physical parameters and are related to the maximum coherency stacking velocity through the r.m.s. velocity. The relationship between these velocities is illustrated. The limitations to the horizontal resolution of the velocity tool are discussed.

The CDP-derived velocities are subject to numerous errors; some errors are generated during the acquisition and processing stages and during the wave propagation in the ground, others arise from geological complexities or are subjective in nature, etc. These errors are discussed and methods of their estimation and, where appropriate, of their treatment are given. The main applications of the CDP-derived velocities are briefly reviewed.

---

† Present address: British Petroleum Exploratie Maatschappij Nederland B.V., Catsheuvel 61, 2617 KA The Hague, Holland.
‡ The term common depth point (CDP) will be used in the text to denote the standard technique whereby the sources and receivers are symmetrically disposed about a common ground point.[4] Strictly, the term is inaccurate when inclined reflectors are involved, but will be adhered to in the text in conformity with common usage.

INTRODUCTION

Historical Note

The possibility of deriving subsurface velocities from seismic reflection data was realised in the early days of reflection techniques. The original methods of velocity estimation were based on a simple model consisting of a horizontal reflector beneath a ground of uniform velocity. In this model, the incident and reflected rays travel along straight line trajectories. The two-way travel-time, T_x, from the source to the reflector and back to a receiver at the same horizontal level as the source, is given by

$$T_x^2 = T_0^2 + \frac{X^2}{V^2} \tag{1}$$

where T_0 is the two-way travel-time corresponding to the normal incidence trajectory, X is the source–receiver offset and V is the propagation velocity which is the velocity being sought. Several methods[1] were designed for estimating this velocity by utilising reflection times corresponding to different source–receiver offsets in accordance with eqn. (1). Gradual elaboration of these early methods led to the development of some useful techniques which were in common use among the pre-CDP generation of geophysicists. The graphical '$T\Delta T$' and 'T^2–X^2' methods were the most widely known. Both methods are described in detail in several books, e.g. Telford et al.[2]

The 'T^2–X^2' method is relevant to subsequent discussions. It is based directly on eqn. (1); if, for a given reflector, T_x^2 values corresponding to a series of X^2 values are plotted on a T_x^2 vs X^2 plot then the inverse of the slope of a straight line fitted through the plotted points gives an estimate of V^2.

Other advances were made in the mid 1950s culminating in Dix's analytical work.[3] Dix considered the case of a reflector at the base of n uniform layers; raypaths corresponding to all offsets other than zero offset suffer refraction at each layer interface, i.e. deviate from a straight line trajectory. Hence, the linear relationship between T^2 and X^2 of eqn. (1) does not hold strictly. At very small offsets where refraction may be regarded as negligible, eqn. (1) is closely approximated. At zero offset ($X^2 = 0$) the inverse of the tangent to the T^2–X^2 plot represents the square of velocity. Dix showed that this velocity is not the true average velocity to the reflector but a time-weighted r.m.s. velocity, $V_{r.m.s.}$, defined in eqn. (3) (see the section on the time and velocity relationships). Dix's analysis also provided a formula for determining the velocity in an interval between two horizons from the travel-time and r.m.s. velocity corresponding to each of them, as will be shown later. A similar approach to that of Dix was being developed independently in Germany at about the same time.[5,6]

Dix's suggested field technique was based on that described by Hansen,[7] which embodied the essential geometric principles of the common depth point techniques. The adoption of these techniques as a standard acquisition practice considerably facilitated the extraction of velocity information from reflection data. At the present we have the benefit of various analytical studies and processing refinements that have been building up during the past two decades. These advances have significantly increased the reliability of the derived velocities and our ability to manipulate them. At the same time, our improved understanding of these velocities has led to an increasing awareness of their many limitations, the accuracy requirements of many applications being frequently greater than can be provided in practice.

Basic Concepts in Velocity Work
In seismic prospecting the term velocity refers to the propagation speed of the seismic wave, a property of the propagation medium. The term velocity analysis is currently used to denote the process of determining velocity from the stacking of CDP data. Often, the term also encompasses a whole suite of subsequent operations used in detailed velocity determinations. The stacking velocity required for velocity studies is that producing maximum coherency in the primary reflection data. Often, this velocity is known simply as the stacking velocity. It is, sometimes, also referred to as the moveout velocity, normal moveout velocity, CDP velocity, etc. None of the above terms is an unambiguous description of the measured velocity. Throughout the text, we shall refer to this velocity as the maximum coherency stacking velocity and denote it by the abbreviated form MCS velocity.

As will be discussed later, the maximum coherency stacking velocity has the dimensions of velocity but is not a physically meaningful quantity. Uses for exploration purposes, apart from data processing, require velocities that are of physical significance, namely the average velocity (defined in the section on average velocity) and the interval velocity (defined in the section on interval velocity). The r.m.s. velocity is a first approximation to the MCS velocity and is simply and directly related to the interval and average velocities. It, therefore, provides an important bridge between the MCS velocity and the interval and average velocities. Figure 1 illustrates schematically the derivation relationship of these velocities. Clearly, Dix's

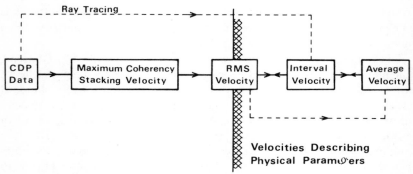

FIG. 1. Schematic illustration of the basic relationships between velocities derived from CDP data.

work, as applied to CDP-derived velocities, has been fundamental in establishing this relationship.

The measured MCS velocity value is inaccurate due to errors generated at the acquisition, processing and other stages of MCS velocity derivation. These errors are transmitted to the estimate of the r.m.s., interval, and average velocities. Further errors are introduced during the derivation of these velocities. The sources of error are summarised in Fig. 2.

The main task of the geophysicist in velocity work is not only to derive meaningful velocities from the MCS velocity but also to investigate the accuracy of the derived velocities in order to assess their validity for the particular purpose for which they are intended. The main uses of CDP-derived velocities are listed in Table 1.

Clearly, velocities derived from direct well measurements cannot be matched in accuracy and resolution by CDP-derived velocities. On the other hand, CDP data provide information regarding the velocity distribution in the ground over the whole survey area and at relatively low cost. In the absence of wells in the area, such information can be extremely useful. The combined CDP and well velocity data, when the latter are available, provide a wide range of possibilities for detailed velocity studies.

Scope of the Present Work

In the present work, velocities obtained from CDP reflection data, currently the main source of velocity information in practice, will be dealt with exclusively. Such specialised topics as velocity determinations linked to section migration,[8,9] velocities from three-dimensional stacking, etc. will not be covered. Work based on reflection coefficients and other studies

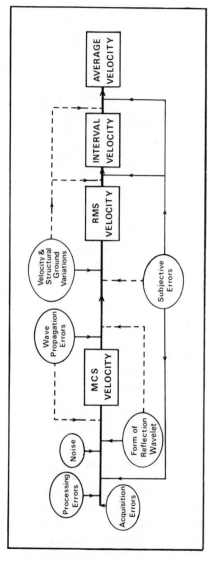

FIG. 2. Main factors affecting the accuracy of derived velocities. Broken lines indicate secondary effects.

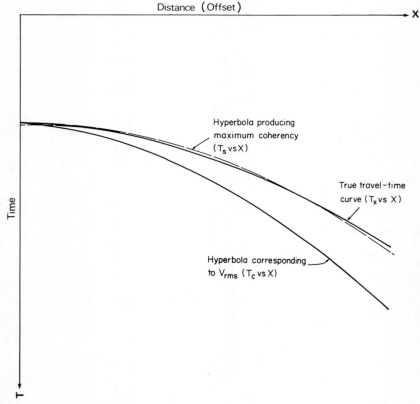

FIG. 3. Simplified illustration of the relation between the true travel-time curve (primary reflection in the CDP gather), the hyperbola producing maximum coherency (corresponding to V_{MCS}), and that corresponding to $V_{r.m.s.}$. The plot is the T–X plane equivalent of a T^2–X^2 plot (cf. Figs. 4 and 5).

relating to data from well measurements also fall outside the scope of this work.

The use of a model consisting of homogeneous isotropic horizontal layers will be implied in the text, except where indicated otherwise. This ideal model is the best understood model in velocity analysis work. Its simplicity helps to illustrate various problems in an instructive manner. The term time–distance curve will be used synonymously with a primary reflection in a CDP record (Fig. 3). The term reflector will be treated as being synonymous with layer interface.

TABLE 1
THE MOST COMMON USES OF CDP-DERIVED VELOCITIES

Velocity	Main uses	Precision requirements
MCS	Stacking of seismic sections	Modest–low
	Preliminary migration processing	Modest–low
	r.m.s. velocity estimation	Dependent on problem
r.m.s.	Estimation of migration velocity	Generally modest
	Interval velocity estimation	Dependent on problem
	Average velocity estimation	Dependent on problem
Interval	Gross lithological and stratigraphical studies	High–modest
	General seismic interpretation purposes	Modest–low
	Age estimation	High–modest
	Detection of zones of abnormal pressure	High–modest
	Ray tracing	Dependent on problem
	Migration processing	Generally modest
	Average velocity estimation	Dependent on problem
Average	Depth conversion	Generally modest
	General seismic interpretation purposes	Modest–low

Precision requirements in terms of equivalent r.m.s. velocity are, very broadly: high = 0·1–1·0%, modest 1–5%, low > 5%. The requirements quoted in the precision column are only intended as a general guide; individual problems vary widely in their requirements.

TIME AND VELOCITY RELATIONSHIPS

The Time–Distance Relationship

The Time–Distance Equation

For a reflector at the base of n uniform horizontal layers, the reflection time, T_x, corresponding to a source–receiver offset, X, is given by[10]

$$T_x^2 = T_0^2 + \frac{X^2}{V_{r.m.s.}^2} + C_3 X^4 + C_4 X^6 + \cdots + C_j X^{2j-2} + \cdots \qquad (2)$$

where the coefficients $C_i (i = 3, 4, \ldots, \infty)$ are functions of the thicknesses and velocities of the n layers; $V_{r.m.s.}$ is the r.m.s. velocity along the normal incidence (zero offset) trajectory defined by

$$V_{r.m.s.}^2 = \frac{\left(\sum_{k=1}^{n} v_k^2 t_k\right)}{T_0} \qquad (3)$$

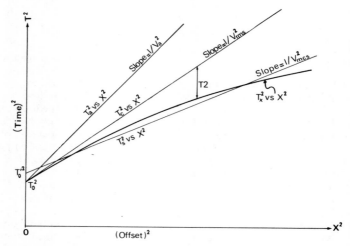

FIG. 4. T^2–X^2 plot showing the relation between the travel-time plot (T_x^2 vs X^2), the straight line best fitting the plot (T_s^2 vs X^2), the tangent to the plot at $X^2 = 0$ (T_c^2 vs X^2), and the straight line corresponding to the refraction-free trajectory (T_a^2 vs X^2).

where v_k and t_k are the velocity and two-way travel-time within the kth layer; T_0 is the zero offset (normal incidence) reflection time, i.e.

$$T_0 = \sum_{k=1}^{n} t_k$$

From eqn. (2), we obtain, at $X^2 = 0$,

$$\frac{d(T_x^2)}{d(X^2)} = \frac{1}{V_{r.m.s.}^2}$$

i.e. $V_{r.m.s.}^2$ is the inverse slope of the tangent to the T_x^2–X^2 plot at $X^2 = 0$, (Figs. 4 and 5). In fact, the inverse slope of any tangent to the T^2–X^2 curve is equal to $V_{R.M.S.}^2$ where $V_{R.M.S.}$ is the r.m.s. velocity for the trajectory corresponding to the X value of the point of tangency.[11] In mathematical notation,

$$\frac{1}{\left[\dfrac{d(T_x^2)}{d(X^2)}\right]_M} = V_{R.M.S.}^2 = \frac{\left(\sum_{k=1}^{n} v_k^2 t_k'\right)}{\sum_{k=1}^{n} t_k'} \tag{4}$$

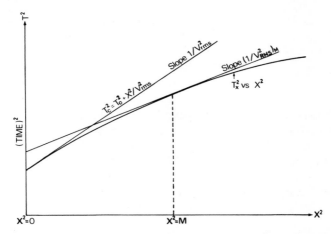

FIG. 5. An illustration of $V_{r.m.s.}$ as a special case of the r.m.s. velocity along the trajectory corresponding to the X value at the point of tangency to the T_x^2 vs X^2 plot.

where t'_k is the two-way transit-time in the kth layer and M is a general point on the X^2-axis (Fig. 5).

The r.m.s. velocity of eqn. (3), inferred by Dix,[3] can now be seen to correspond to a special (normal incidence) case of the general relationship of eqn. (4).

Hyperbolic Relationships

As shown by eqn. (2), the general relationship between T_x^2 and X^2 is not linear, that is, the time–distance curve (T_x vs X) is not hyperbolic (Figs. 3 and 4); a truly hyperbolic time–distance relationship arises only in the ideal case of a one-layer model, i.e. refraction-free raypath, in which case eqn. (2) reduces to eqn. (1).

However, current velocity analysis algorithms are based on hyperbolic time–distance relationships, largely due to implementation simplicity. The MCS velocity, V_{MCS}, corresponds to the hyperbola producing maximum coherency (Fig. 3). In terms of the T^2–X^2 plot of Fig. 4, V_{MCS}^2 is, effectively, given by the inverse slope of the best fit straight line through the true travel-time plot (T_x^2 vs X^2). This straight line has the form

$$T_s^2 = T_0'^2 + \frac{X^2}{V_{MCS}^2} \tag{5}$$

the intercept $T_0'^2$ being slightly different from T_0^2.

Equation (5) is a hyperbolic approximation to the time–distance relationship and not a hyperbolic truncation which, from eqn. (2), would give

$$T_c^2 = T_0^2 + \frac{X^2}{V_{r.m.s.}^2} \qquad (6)$$

For this reason, V_{MCS} does not equal $V_{r.m.s.}$. The common use of V_{MCS} as a first approximation to $V_{r.m.s.}$ is equivalent to using the inverse slope of the T_s^2 vs X^2 line as an estimate for the inverse slope of the T_c^2 vs X^2 line in Fig. 4. The relationship between V_{MCS} and $V_{r.m.s.}$ is discussed further in the section on the relation between MCS and r.m.s. velocities.

For the straight-line (refraction-free) incident and reflected trajectories the hyperbolic relationship

$$T_a^2 = T_0^2 + \frac{X^2}{V_a^2}$$

is obtained, where V_a is the average velocity to the reflector defined in eqns. (15) and (16). The principle of least time path requires that $T_a > T_x$ for $X \neq 0$ (Fig. 4).

Convergence Properties of Eqn. (2)

Two-term truncation. The discrepancy between eqn. (2) and its two-term truncation in eqn. (6) is given by

$$T2 = T_c^2 - T_x^2 = \sum_{j=1}^{\infty} p^{2j} F_j(v, h) \qquad (7)$$

If the thickness, velocity and angle of incidence pertinent to the kth layer are denoted by h_k, v_k and θ_k respectively and the depth of the reflector at the base of the nth layer by D then: $p(=\sin\theta_k/v_k)$ is the ray parameter; $F_j(v, h)$ is a complicated non-negative function of the velocities and thicknesses of the layers above the reflector, increasing in magnitude as the quantity

$$g = \frac{1}{D^2} \sum_{i=1}^{n-1} h_i \sum_{k=1}^{n} h_k \frac{(v_i - v_k)^2}{v_i v_k} \qquad (8)$$

increases; and g is a non-negative quantity that gives a measure of the velocity heterogeneity of the ground above the reflector. It will be referred to as the velocity heterogeneity factor. Equation (7) shows that $T2 = 0$ in the case of a one-layer model ($g, F_j = 0$) and at normal incidence ($X, p = 0$). In all other cases, $T2 > 0$.

For non-normal incidence, the larger is the velocity heterogeneity ($\equiv F_j$), the greater is the deviation of the raypath from a straight line trajectory, i.e. the greater is the amount of refraction. Also, increasing offset–depth ratio ($\equiv p$), increases refraction. Therefore, the physical interpretation of eqn. (7) is, simply, that the greater is the refraction along the raypath the larger is the amount by which T_c (refraction-free) exceeds the true travel-time. The implications of eqn. (7) are discussed further in the section on the relation between MCS and r.m.s. velocities.

Higher terms. C_3 in eqn. (2) is a negative coefficient.[10,12] Hence, since a two-term truncation overestimates T_x^2, the inclusion of the negative third term will improve the convergence of the series of eqn. (2) unless

$$C_3 X^4 > 2(T_c^2 - T_x^2)$$

in which case the truncated series is over-corrected and the result will be less accurate than truncating at two terms. In practice, $C_3 X^4$ is almost always very close to $T_c^2 - T_x^2$ so that a three-term truncation produces usually very accurate results.[12]

A study of C_4 and higher coefficients shows that the inclusion of further terms beyond the third term will not necessarily improve the convergence. Strong oscillations can occur as these terms are added, especially at large offset–depth ratios.[12]

The Normal Moveout

The normal moveout (NMO) may be defined as the shift that must be applied to a reflection time at offset X to reduce it to the time that would have been recorded at zero offset (normal incidence). Accordingly, the NMO would be given by

$$\Delta T = T_x - T_0$$

In MCS velocity determinations, several stacking velocities are attempted. For each stacking velocity, V_s, NMO corrections are applied to the traces of the gather according to

$$\Delta T = \left(T_0'^2 + \frac{X^2}{V_s^2} \right)^{1/2} - T_0' \tag{9}$$

or a similarly convenient relationship.

The NMO is frequently encountered in the literature[13] in its less rigorous form

$$\Delta T = T_c - T_0 = \left(T_0^2 + \frac{X^2}{V_{r.m.s.}^2}\right)^{1/2} - T_0$$

and, sometimes, in its approximate parabolic form

$$\Delta T = T_c - T_0 \simeq \frac{X^2}{2T_0 V_{r.m.s.}} \tag{10}$$

Such forms were also the basis of the graphical $T\Delta T$ method of velocity analysis.

MCS and r.m.s. Velocities

General Remarks

In obtaining MCS velocities from the CDP gather, a succession of NMO corrections are applied according to the scheme of eqn. (9) or some parabolic equivalent. For a given primary reflection, the value of V_s producing maximum coherency in the corrected traces is the MCS velocity.

The coherency measure varies from one velocity analysis algorithm to another. Ignoring these variations, we may closely simulate the process of obtaining MCS velocity by a least squares fit of the T_s^2 vs X^2 line to the T_x^2 vs X^2 curve of Fig. 4. The simulation gives

$$V_{MCS}^2 = \frac{\left[m \sum_{i=1}^{m} X_i^4 - \left(\sum_{i=1}^{m} X_i^2\right)^2\right]}{\left[m \sum_{i=1}^{m} T_i^2 X_i^2 - \left(\sum_{i=1}^{m} T_i^2\right)\left(\sum_{i=1}^{m} X_i^2\right)\right]} \tag{11}$$

where m is the number of stacked traces and T_i and X_i are the offset and true travel-time corresponding to the ith trace.

Lack of Physical Significance of MCS Velocities

As MCS velocity determinations are based on hyperbolic time–distance relationships, the measured V_{MCS} only represents the true propagation velocity in the trivial case of a one-layer model. In the multi-layer case, the physical significance of V_{MCS} becomes ambiguous as a consideration of eqns. (5) and (11) would show. However, the determination of $V_{r.m.s.}$ from V_{MCS}, in this case, entails only the removal of the difference between the two quantities through simple procedures (see the section on the relation between MCS and r.m.s. velocities, and the section on refraction). In practice, where conditions are much more complicated than in the above ideal models, V_{MCS} is rendered a mere mathematical parameter applied in eqn. (9) to produce optimum coherency in the stacked traces. A correct estimation of $V_{\text{r.m.s.}}$ from V_{MCS} involves many corrections which will be discussed in the section on factors affecting velocity determinations. For the present purpose, the relationship between V_{MCS} and $V_{\text{r.m.s.}}$ will be illustrated in terms of the ideal multi-layer model.

Relation Between MCS and r.m.s. Velocities
As shown by eqn. (7), $T_c > T_x$ at non-normal incidence. Hence, V_{MCS}, which is based on a best fit to the true travel-time curve, will invariably overestimate $V_{\text{r.m.s.}}$. Under ideal, noise-free conditions, the quantity

$$B = V_{\text{MCS}} - V_{\text{r.m.s.}} \qquad (12)$$

represents the bias in the estimate. The magnitude of B depends on the magnitude of $T2$ (see the section on convergence properties of eqn. (2)) corresponding to each trace. Therefore, B increases with increasing ground heterogeneity factor, g, and ray parameter, p, i.e. with increasing refraction corresponding to each trace. In fact, $V_{\text{r.m.s.}}$ may be regarded as the limit to which V_{MCS} tends as the spread length (hence, refraction) diminishes. For a theoretical spread of infinitesimally small length, the slopes of the T_c^2 vs X^2 and T_s^2 vs X^2 lines coincide, thus, giving

$$(V_{\text{MCS}}^2)_{X^2=0} = V_{\text{r.m.s.}}^2$$

Over the same ground, p increases with increasing trace offset and decreases with increasing reflector depth. Therefore, there is always a rapid increase in the bias as the spread length increases. However, for the same spread geometry, the bias does not necessarily diminish with increasing reflector depth;[14] a large increase in g at depth and the consequent increase in F (v, h) could swamp the decrease in p and cause B to increase (see the examples in the next section).

The bias can be expressed in a closed form[15,16] through the following approximate equation:

$$B \simeq \frac{\left[\left(\sum_{k=1}^{n} v_k^4 t_k\right) - V_{\text{r.m.s.}}^4 T_0\right] X_e^2}{8 V_{\text{r.m.s.}}^5 T_0^3} \qquad (13)$$

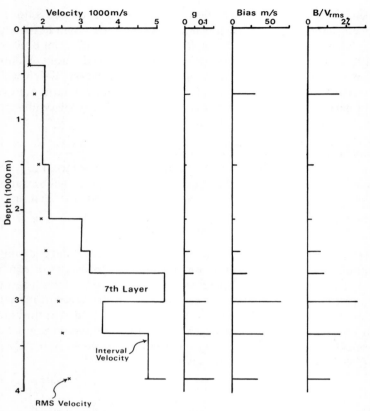

FIG. 6. Variation of the bias in the estimate of $V_{r.m.s.}$ with depth for a simplified model from the North Sea.[14] Maximum offset = 3000 m.

where X_e is an effective offset given by

$$X_e^2 = \frac{\sum_{i=1}^{m} X_i^2}{m} \qquad (14)$$

X_i is the offset of the ith trace and m is the number of stacked traces.

Examples

An example of the variation of the bias with depth is shown in the simplified model of Fig. 6; the heterogeneity factor in the top four layers is low. Consequently, the bias decreases steadily with depth as the spread

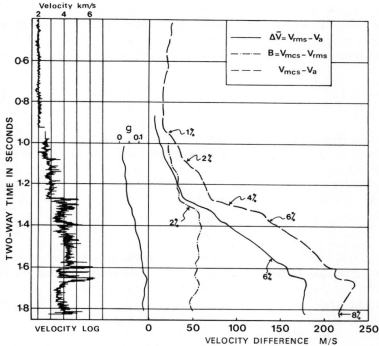

FIG. 7. Differences between average, r.m.s. and MCS velocities. Maximum offset = 3000 m. Modified from O'Brien and Lucas.[34]

length–depth ratio gets smaller. The large change in the velocity of the seventh layer produces a large increase in the heterogeneity factor. Thus, the bias increases significantly at the base of the seventh layer. Below that, the increase in the heterogeneity factor is insufficient to compensate for the effect of decreasing p with depth and the bias decreases downwards. A second example is shown in Fig. 7. The variations in B can be similarly explained in terms of the velocity heterogeneity in the ground.

Average Velocity
General Remarks

The average velocity along the source–reflector–receiver trajectory is given by the general form

$$V_a = \frac{1}{T} \int_0^T v_{\text{ins}}(t)\, dt \qquad (15)$$

where v_{ins} is the instantaneous velocity, defined as the velocity in an infinitesimally small interval, and T is the total travel-time. In particular, in a horizontally layered ground, the average velocity to the base of the nth layer along the normal incidence trajectory is given by

$$V_a = \frac{1}{T_0} \sum_{k=1}^{n} v_k t_k \qquad (16)$$

The corresponding time-weighted r.m.s. average velocity (or simply the r.m.s. velocity) is defined in eqn. (3), in the section on the time–distance equation.

Other forms of weighted average velocities are sometimes encountered in the literature, e.g. the generalised form

$$V_t^n = \frac{1}{T} \int v_{ins}^n(t)\, dt$$

In order to avoid ambiguities arising from the occasional use of the term average velocity to denote various forms of velocity averaging, we shall restrict the use of the term to that defined by equations (15) and (16), especially the latter.

Relation Between Average and r.m.s. Velocities
The average and r.m.s. velocities are related by[14]

$$\frac{V_{r.m.s.}}{V_a} = (1 + g)^{1/2} \qquad (17)$$

g being the heterogeneity factor defined in eqn. (8). Another useful relation is through the quasi anisotropy factor, A_0,

$$\frac{V_{r.m.s.}}{V_a} = A_0 \qquad (18)$$

where

$$A_0^2 = 1 + \frac{\sigma_v^2}{V_a^2}$$

σ_v is the standard deviation of the velocity distribution computed by dividing the ground into a series of layers of equal transit time.[17] A_0 is related to g in

$$g = A_0^2 - 1$$

Equations (17) and (18) show that, in a multi-layer ground, $V_{r.m.s.}$ invariably exceeds V_a, the difference between the two quantities being dependent on the velocity heterogeneity in the ground.

An example of the variation of $\Delta \bar{V}(= V_{r.m.s.} - V_a)$ with depth is shown by the middle curve in Fig. 7. The curve is representative of the order of values to be expected normally. Note that the sudden increase in interval velocity at about 1·26 s produces a corresponding increase in the heterogeneity factor and, consequently, a steeper rise in $\Delta \bar{V}$ down the well. The reverse effect is observed at 1·65 s.

The direct determination of average velocity from r.m.s. velocity requires interval velocity information for use in eqn. (17) or (18). In practice, if a velocity log is available from a well in the area, a percentage difference between the r.m.s. and average values can be derived for each reflection. This percentage may then be applied over the whole area providing, of course, that the validity of such application is not lost through drastic structural, lithological, or other changes.

Interval Velocity

The traditional definition of interval velocity is that it is the average velocity over an interval between two depths. This definition is valid when the interval velocity is derived from average velocities (e.g. from well data):

$$v_{int} = \frac{(v_b T_b - v_a T_a)}{(T_b - T_a)} \quad (19)$$

where v_{int} is the interval velocity, v_a and v_b are the average velocities at the top and bottom of the interval and T_a and T_b are the corresponding normal incidence times.

With velocities derived from reflection data, the interval velocity is computed from r.m.s. velocities through the equation of Dix:[3]

$$v_{int} = \left[\frac{(V_b^2 T_b - V_a^2 T_a)}{(T_b - T_a)} \right]^{1/2} \quad (20)$$

where V_a and V_b are the r.m.s. velocities at the top and bottom of the interval. From an analysis of eqn. (20), it can be shown that the interval velocity thus calculated is the r.m.s. (not average) velocity of the interval.[14] The above results can, therefore, be summed up as follows:

An interval velocity derived from average velocities (eqn. (19)) is the average velocity of the interval;

An interval velocity derived from r.m.s. velocities (eqn. (20)) is the r.m.s. velocity of the interval.

In interval velocity measurements from CDP data, the intervals are, in most cases, selected between horizons of some geological significance. Individual units within such intervals usually have fairly similar velocities. Any bands of markedly different velocity are often very thin and do not produce a significant rise in the velocity heterogeneity factor of the interval. In these cases, the measured (r.m.s.) interval velocity will be, generally, close to the average velocity value of the interval. When the interval consists of very varied units a correction factor, based on velocity logs from wells in the area (as described in the section on average velocity), can be used for estimating the average interval velocity from the measured interval velocity. The discrepancy between average and r.m.s. velocities in individual intervals should also be taken into account when computing the average velocity to a reflector through eqn. (16).

Dipping Reflectors

It has been shown by Levin[18] that for a plane dipping reflector below a uniform overburden of velocity V

$$T_x^2 = \frac{[4D^2 + X^2(1 - \sin^2 \phi \cos^2 \theta)]}{V^2} \quad (21)$$

where D is the length of the normal to the reflector from the midpoint between the source and the receiver, ϕ is the true dip, and θ is the angle between the direction of the true dip and the profile. If the dip component in the direction of the profile is α then, from eqn. (21),

$$T_x^2 = T_0^2 + \frac{X^2 \cos^2 \alpha}{V^2} \quad (22)$$

where T_0 is the zero offset travel-time. Because of this hyperbolic time–distance relationship the stacking process will not detect the presence of dip. The resulting MCS velocity will be, simply,

$$V_{\text{MCS}} = \frac{V}{\cos \alpha} \quad (23)$$

Equation (23) is the two-dimensional form for a single reflector.

Shah[19] uses wavefront curvature to treat the general two-dimensional case of n reflectors of arbitrary dips separating layers of uniform velocity. The main result is the relationship,

$$V_{\text{NMO}}^2 = \frac{1}{T_0 \cos^2 E_0} \sum_{k=1}^{n} v_k^2 t_k \prod_{j=0}^{k-1} \frac{\cos^2 I_j}{\cos^2 E_j} \quad (24)$$

with
$$\frac{\cos^2 I_0}{\cos^2 E_0} = 1$$

where V_{NMO}^2 is defined as the inverse slope of the tangent to the $T_x^2-X^2$ plot at $X^2 = 0$, i.e. V_{NMO} is the limit to which V_{MCS} tends at zero offset (normal raypath)—cf. $V_{r.m.s.}$ in the case of horizontal layering; v_k and t_k are, respectively, the interval velocity and the transit time within the kth layer; I_j and E_j are the angles of incidence and emergence for the jth reflector; and T_0 is the total travel-time. The last four quantities refer strictly to normal incidence trajectories.

The above relationship is very similar to that obtained by Dürbaum[5] for single cover geometry. Equation (24) is of the same type as eqn. (3) and reduces to it in the absence of dip. V_{NMO} can be estimated from the MCS velocity (see the section on plane dipping reflectors), but, for short spreads, the direct approximation $V_{NMO} \simeq V_{MCS}$ is usually valid.

From eqn. (24) it is possible to estimate the interval velocities and the positions of the layer interface. The estimation is carried out by stripping successive layers in the following way:[19] let us assume that the positions of the top $n - 1$ reflectors (layer interfaces) and the intervening interval velocities have been determined, then the emergence angle E_0 corresponding to the nth reflector is obtained from

$$\sin E_0 = 0.5 v_1 S_n \qquad (25)$$

where S_n is the time dip (in ms/km) of the nth reflector as obtained from the stacked seismic section. The emergent ray is then traced to the $(n - 1)$th reflector. With t_j, E_j and $I_j (j = 1, 2, \ldots, n - 1)$ determined, t_n is found by subtracting all the t_j's from T_0. v_n is then found by substitution in eqn. (24). The determination of the position of the nth reflector follows from elementary geometrical principles. There are special cases where the ray tracing stage becomes unnecessary. The single reflector model is an obvious case. In the case of n parallel reflectors eqn. (24) reduces to

$$V_{NMO} = \frac{V_{r.m.s.}}{\cos E_0}$$

which is a similar result to that obtained by Dix,[3] $V_{r.m.s.}$ being the r.m.s. velocity along the normal incidence raypath. The interval velocities can be obtained from eqn. (20).

The three-dimensional case of arbitrarily dipping plane interfaces has been dealt with by Krey[20] and Hubral.[21,22] As in the two-dimensional case, the treatment is essentially based on the normal raypath trajectory. The estimation of the interval velocities and reflector positions requires the provision of estimates of V_{NMO} and the time dip of each reflector along the profile (from the seismic section) and in some other direction, preferably, at right angles to the profile. The approach parallels that of Shah.[19] The value of v_n is determined from a solution of an equation relating v_n to V_{NMO} through a number of matrix functions. The mathematical treatment is clearly presented by both authors but is too long to describe here. The reader is referred to the original works for the details.

CURRENT TECHNIQUES

Basic Principles

As mentioned earlier, velocity analysis techniques are based on the assumption of straight-line (refraction-free) raypaths. Hence, the basic scheme in velocity analysis consists of performing a stack across the CDP gather along hyperbolic trajectories defined by eqn. (9). To illustrate the principle of these techniques let us consider a hypothetical CDP gather (Fig. 8). Suppose that the noise-free reflection in Fig. 8 forms an exact hyperbola, H_{op}, and that the zero offset time corresponding to the peak of the reflection is t_0. Suppose also that the velocity analysis is to be carried out with reference to t_0 and that the range of velocity to be covered by the analysis is V_a to V_b. The analysis is carried out as follows:

(1) An initial stacking velocity $V_1 = V_a$ is assumed. This velocity corresponds to hyperbola H_1. NMO corrections, computed from eqn. (9), are then applied. This is equivalent to aligning the traces according to hyperbola H_1.
(2) The degree of match (or coherency) between the traces at this alignment is measured, for example, by summing the amplitudes at t_0 and determining the output power, i.e. the absolute value of the summation.
(3) The velocity is then incremented by an appropriate step and new NMO corrections are applied. The coherency is again measured.
(4) Step (3) is repeated until V_b is reached.
(5) The zero offset time is then incremented from t_0. Steps (1)–(4) are repeated.
(6) The above process is repeated until the appropriate range of time down the CDP record has been covered.

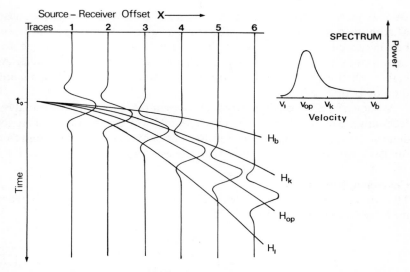

FIG. 8. Hypothetical example showing a series of hyperbolic trajectories. Hyperbola H_{op} produces the highest coherency at zero offset time t_0. The spectrum for t_0 is shown on the right.

In practice, the amplitude summation (or any other coherency measure) is carried out within a time gate, usually centred about the reference zero offset time. Thus, in a gate of width t_g, centred about t_0, the coherency is based on samples between $t_0 - 0.5 t_g$ and $t_0 + 0.5 t_g$. The gate width is generally of the order of 0·75–1·5 times the predominant wavelet period. There is, usually, adequate overlap between successive gates.

In many algorithms, it is the NMO, not the stacking velocity, that is incremented in step (3). As indicated by the form of eqn. (10), the reflection wavelets are stretched by the application of NMO corrections at a rate that increases with increasing offset and diminishes down the record. This stretching produces wavelet distortions which affect the accuracy of the measured velocity as explained in the section on offset-related wavelet changes.

The results of analysis in steps (1)–(4) can be displayed as a plot of coherency measure (power in the example) vs velocity. Such a plot is known as a velocity spectrum.[10,23] A new spectrum is produced at every time increment so that the analysis of the whole CDP record is displayed as a series of these spectra. The spectrum corresponding to t_0 is shown in Fig. 8. This spectrum illustrates the increase in power build-up as the stacking velocity increases from V_1 and the traces are gradually brought in phase. Maximum power (maximum coherency) is attained when all the traces become exactly in phase, i.e. when the stacking velocity reaches a value that corresponds to the hyperbola H_{op}. This is the MCS velocity value. As higher velocities are applied, the traces begin to get out of phase again and the power diminishes.

As shown by the preceding hypothetical example, current velocity analysis techniques involve three main processes: NMO corrections, coherency determinations, and display of results. The NMO corrections require no further treatment, but the other two processes will now be discussed.

Coherency Measures
The amplitude summation approach[24] is a simple form of coherency measure. It can be represented by

$$C_a(t_0, V_p) = \frac{\sum_{j=r}^{r+q} \left| \sum_{i=1}^{m} A_{ij} \right|}{(q+1)} \qquad (26)$$

where t_0 is the reference zero offset time, V_p is the assumed velocity, $(q+1)$ is the number of samples within the time gate, A_{ij} is the amplitude of the jth sample on the ith trace after NMO application, m is the number of traces, and $r = t_0 - q/2$.

Cross correlation is often used as a coherency measure. Some velocity analysis algorithms employ an unnormalised cross correlation measure. An early application of the unnormalised cross correlation was the dynamic correlation method of Schneider and Backus.[13] The method is based on residual moveout analysis after an initial NMO correction and relies on averaging results from adjacent ground points to enhance accuracy.

Normalised cross correlation is also a widely used coherency measure. Neidell and Taner[25] distinguish between a statistically normalised and an energy normalised cross correlation function. The latter has the general form

$$C_N = \frac{\sum_{j=r}^{r+q} \left(\sum_{i=1}^{m} A_{ij(i)} \right)^2 - \sum_{i=1}^{m} A_{ij}^2}{(m-1) \sum_{j=r}^{r+q} A_{ij(i)}^2} \qquad (27)$$

The energy normalisation is preferred to the statistical normalisation owing to its power to penalise amplitude variations in the input traces and because of its computational superiority.

The semblance coefficient[10,25] is a normalised output–input energy ratio defined by

$$S_c = \frac{\sum_{j=r}^{r+q} \left(\sum_{i=1}^{m} A_{ij(i)} \right)^2}{m \sum_{j=r}^{r+q} \sum_{i=1}^{m} A_{ij(i)}^2} \qquad (28)$$

From eqns. (27) and (28) we obtain

$$S_c = \frac{1 + C_N(m-1)}{m}$$

A third approach in coherency measures is through spectral techniques (not to be confused with displays of velocity spectra). Robinson[26] has proposed a frequency–wavenumber domain method which is equivalent to the dynamic correlation analysis method. However, this approach has not gained popularity due to a number of operational difficulties. Coherency measures employed in the automatic picking of seismic sections[27,28] are predominantly based on semblance or amplitude summation criteria.

It should be noted that, despite differences in detail, the philosophy behind the various coherency criteria is essentially the same. Though no two velocity analysis techniques working on the same data would produce exactly identical results, the differences are usually minor. The amplitude summation criterion has occasionally been favoured on account of its implementation simplicity and, sometimes, the good accuracy obtainable under certain conditions of low signal–noise ratio. In most other instances, cross correlation criteria are preferred because of their higher accuracy under general practical conditions. The semblance criterion is currently quite widely used.

Display of Velocity Analysis Results
Velocity Spectra Displays
Velocity spectra are the most common form of displaying the results of velocity analysis. In these displays, the coherency value as a function of time and velocity is represented on plots with time varying along one axis and velocity varying along the other. These displays are frequently accompanied by numerical tables of the measured coherency values for verifying the MCS velocity estimate from the display.

Figure 9 shows a variable density display. The density varies in steps; each step encompasses a range of coherency values. A number of distinct

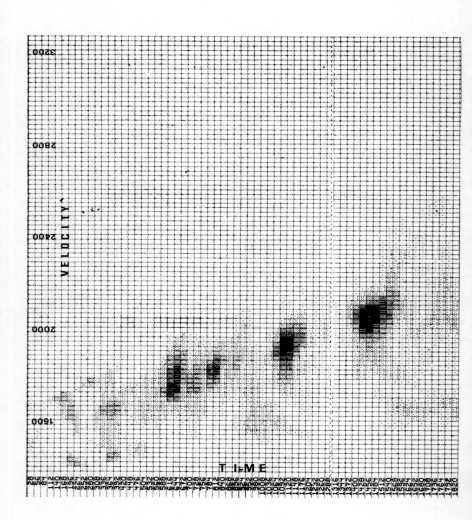

VELOCITY DETERMINATION FROM SEISMIC REFLECTION DATA 25

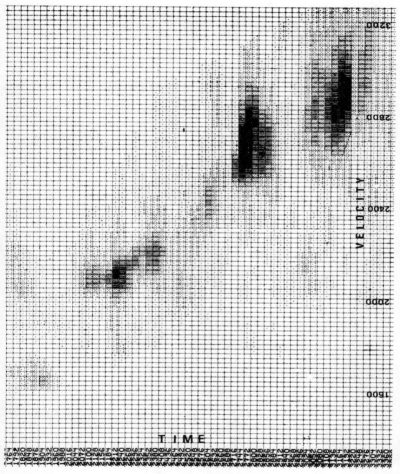

FIG. 9. Variable density display of velocity analysis results. The output is at 20 m/s interval of velocity and 28 ms interval of t_0. The MCS velocity increases from 1800 m/s at about 640 ms to 2860 m/s at about 3170 ms.

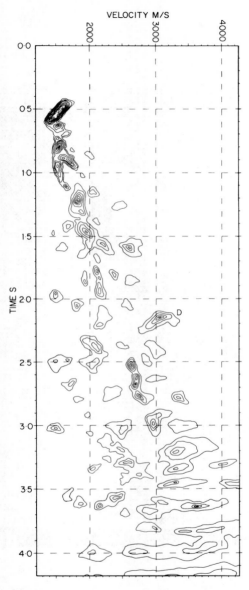

Fig. 10. Contoured velocity spectra display. Below 2 s there is some deterioration in quality caused by a low signal–noise ratio and interference by spurious events. The build-up denoted by D is probably due to a diffraction.

high coherency build-ups can be seen. Each of these build-ups corresponds with a strong primary reflection on the seismic section. At time 1·456 s, the maximum coherency occurs at a velocity of 2060 m/s. In this example, the quality of the CDP data and, consequently, the quality of the display is much better than is normally encountered in practice. Figure 10 illustrates another type of display consisting of contours of equal coherency (semblance) values.

General Features and Interpretation of Velocity Spectra Displays

The interpretation of a velocity spectra display is basically a process of identifying coherency build-ups corresponding to primary reflections and isolating those due to multiples and other spurious coherent events. In carrying out this process, it is important to maintain close reference to the seismic section and the CDP record display, when available. In this way, the presence and strength of primary reflections at the appropriate travel-times can be established. Also the presence of non-primary events can be checked. Conversely, the recognition of non-primary coherency build-ups on the velocity display can be a valuable guide for the processing and interpretation of the seismic section.

The interpretation of the display of Fig. 9 is very simple as there are no prominent non-primary coherency build-ups. In Fig. 10, the primary coherency build-ups are readily recognisable down to about 2 s. Below that, the interpretation is difficult without close reference to the seismic section. A useful quick test in such cases is to verify that the picked velocities produce plausible interval velocities when treated as r.m.s. velocities in eqn. (20).

It can be seen in Fig. 10 that the contours become increasingly stretched, along the velocity axis, with increasing travel-time. This stretching represents a deterioration in the sensitivity of the coherency measure with increasing time and causes a loss in resolution. It results from the fact that, with increasing time, large stacking velocity changes correspond to increasingly smaller NMO changes (cf. eqn. (10)) and low frequencies become increasingly predominant. The display of Fig. 10 shows a large number of coherency build-ups besides those due to primary reflections. Many of these build-ups correspond to multiple reflections. Depending on the details of the multiple trajectory, a multiple could indicate a higher or a lower velocity than a primary of a similar zero offset time. Generally, however, low velocity multiples are more common (see the section on multiples).

Diffractions are another important cause of non-primary coherency build-ups. In the absence of steep dips and complex structures, a diffraction usually indicates a higher (often much higher) velocity than a primary occurring at a similar time (see the section on diffractions). The build-up denoted by D, in Fig. 10, is possibly due to a diffraction.

Cycle skipping (reflection miscorrelation) often gives rise to spurious coherency build-ups which can be quite strong specially in noisy records. These build-ups may indicate a higher or a lower velocity than a primary of a similar time. Cycle skipping is more liable to occur at the early part of the record where high frequencies are abundant and the NMO is large. Sideswipes are another source of spurious coherency build-ups particularly in structurally complex areas.

Miscellaneous Forms of Display

Besides velocity spectra displays, many forms of less general velocity analysis displays have continually been introduced since the early 1970s. A number of display types that are currently among the most common will now be reviewed.

In the 'dip-bar' display,[28] automatically picked reflection wavelets are displayed as short dipping bars. Each bar is annotated with the numerical value of the estimated MCS velocity. This form of display provides readily available continuous velocity data from which the appropriate velocity profiles can be constructed.

Some types of display are intended to provide visual aid for selecting velocities for processing purposes. These displays are of little use for other purposes because of the coarseness of the velocity increment between successive displays. Constant velocity stacks are a common form of such displays. They are produced by stacking the same section or part of section a number of times. In each stack, a single stacking velocity is used throughout. Reflections will stack up with varying strength according to the applied velocity. The appropriate stacking velocities for enhancing primaries and suppressing multiples and other unwanted events are then determined from displays of the stacked sections.

In the display type of Fig. 11, velocity spectra corresponding to a particular reflection are produced at successive shot-point locations along the seismic profile. The display serves as a continuous velocity profile. When spectra for several horizons are displayed together, lateral velocity variations that are common to some or all of the horizons will become apparent.

Colour has been in use for some time to display velocity information directly on the seismic section. Each colour encompasses a velocity range

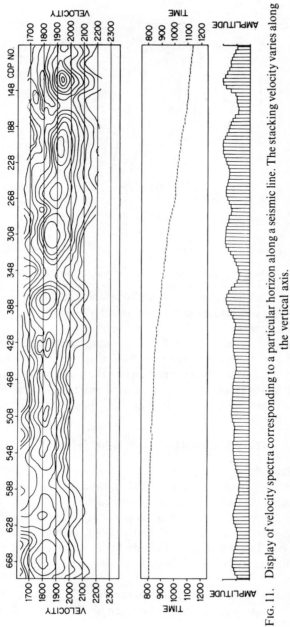

FIG. 11. Display of velocity spectra corresponding to a particular horizon along a seismic line. The stacking velocity varies along the vertical axis.

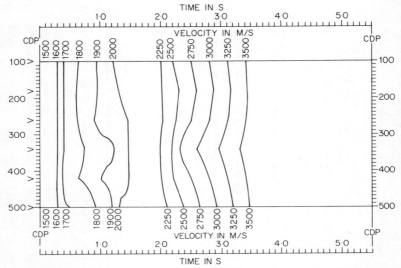

FIG. 12. Contours of equal average velocity values along a seismic line.

according to a specific scheme. In MCS velocity displays, the colours are assigned to the reflections. In interval velocity displays, the colours are assigned to the intervals.

Many facilities currently exist for producing contoured displays of interval, MCS, r.m.s., and average velocities. Figure 12 shows an average velocity contoured display. When produced at the same scale as the seismic section, these displays can be superimposed on the section to convey velocity information directly, in the same way as a colour display.

LATERAL RESOLUTION AND STATICS

Preliminary Remarks

Subsurface inhomogeneities produce different time shifts in different traces of the CDP gather according to the details of each raypath trajectory. The most common source of such time shifts is variations in the near-surface, such as those due to variations in the sea-bed topography or in the weathering layer. Deeper sources include facies changes, salt and shale masses, local gas accumulations, structural changes, ancient topography, etc.

These time shifts will be referred to as statics, but a given trace need not be shifted by the same amount along its entire length. The use of the term will not coincide with its general usage in data processing. The term delays (positive or negative) will also be used to describe the time shifts. The term wavelength will be used loosely to describe the length of the predominant component of repetitive variations or oscillations.

Statics which vary rapidly within the traces of the CDP gather are effectively random noise. They are averaged out in the stacking process and, therefore, produce only small inaccuracies in the measured MCS velocities. Statics varying with a wavelength which is many times greater than a spread length act approximately as a constant time shift. The reflection curvature in the CDP gather remains essentially unaltered. The measured MCS velocity will be close to the correct velocity. Statics varying with a wavelength which is commensurate with a spread length significantly distort the reflection curvature in the CDP gather. The NMO of the 'best fit hyperbola' acquires rapidly varying magnitudes between successive velocity analysis points. Consequently, a plot of MCS velocities, for a given reflector, will oscillate along the seismic profile, as commonly observed in practice. The oscillations have a predominant wavelength which is close to a spread length.

The above considerations show that lateral resolution along the velocity profile depends on the length of the spread. Whether a given velocity variation will escape detection, cause the MCS velocity profile to 'resonate', or be correctly detected depends on whether the predominant wavelength of the time shifts generated by such variations is much shorter than, comparable with, or much greater than a spread length. From the viewpoint of velocity analysis, variation producing the first two types of time shifts are velocity 'anomalies'. The third type is a 'genuine' velocity variation. Clearly, there is complete gradation from one type of variation to another.

It should also be noted that scale is an important factor in lateral resolution. A velocity variation which acts as a velocity anomaly in a problem involving a long spread might behave as a genuine velocity variation in a problem involving a much shorter spread.

The problem of statics will now be presented in terms of simplified models. The cases of shallow and deep statics will be dealt with separately. Methods of correcting statics-generated errors are discussed in the section on laterally variable time delays (statics).

Models of Near-Surface Statics
General Assumptions
The general model consists of a horizontal reflector beneath uniform

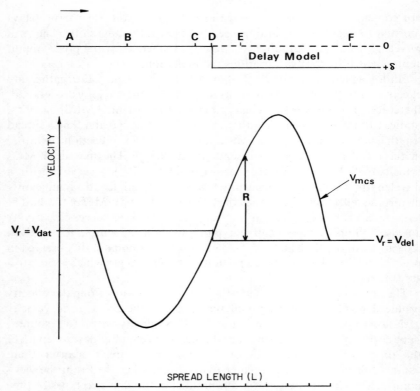

FIG. 13. The response curve across a vertical step.

overburden of velocity V_{dat}. Refraction-free ray trajectories are assumed throughout. The delays are assumed to originate over an infinitesimally small thickness so that no allowance need be made for the angle at which the downgoing and upgoing rays meet the surface.

The response of the velocity analysis to the presence of delays is defined as the difference between the measured MCS velocity, V_{MCS}, and the true average velocity to the reflector, V_r:

$$R = V_{MCS} - V_r \qquad (29)$$

The Step Model
Consider a feature, just below the surface, with its edge at point D, producing constant delays of magnitude δ (Fig. 13). A spread in a CDP

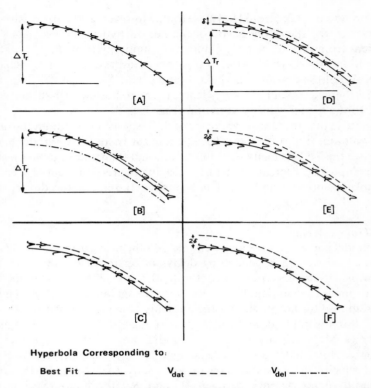

FIG. 14. Diagrammatic illustration of the change in the best fit hyperbola as the CDP gather moves across a vertical step.

configuration moves in the direction AF. Below the feature producing the delays, $V_r = V_{del}$; elsewhere, $V_r = V_{dat}$. Because of equal time shifts, the NMO corresponding to V_{del} is equal to that corresponding to V_{dat}. This NMO will be denoted by ΔT_r.

Figure 14 illustrates various stages of distortion in the reflection curvature as the spread moves across point D. The corresponding response curve is shown in Fig. 13. Note that, as a result of distortions, the response curve is of much greater magnitude than the actual velocity variation ($V_{dat} - V_{del}$). Position B illustrates the effect of distortions clearly. If only the inner traces were stacked, the best fit hyperbola would coincide with that corresponding to V_{dat}, i.e., V_{MCS} would be equal to V_r. If only the outer (delayed) traces were stacked, the NMO of the best fit hyperbola would be equal to ΔT_r and V_{MCS} would fall close to V_r. However, as the entire gather is involved in the stack, the NMO of the best fit hyperbola is significantly different from ΔT_r so that V_{MCS} differs substantially from V_r.

Positions (B)–(E) (Fig. 14) show that the curvature distortion is such that the NMO of the best fit hyperbola is larger than ΔT_r when $V_r = V_{dat}$ and smaller than ΔT_r when $V_r = V_{del}$. Hence, the response varies in the opposite sense to the actual velocity variation (Fig. 13).

Note also that the (anti-) symmetry of the response curve about the point D is not exact (Fig. 13). This is because a negative increment in the NMO increases the MCS velocity by a greater amount than an equal but positive increment would decrease it. Thus, averaging the response curve over, for example, a spread length, does not completely remove the discrepancy between V_{MCS} and V_{dat}.

The Trough Model

This model consists of a velocity anomaly in the near-surface bounded by two inclined sides and generating delays of up to 10 ms (Fig. 15). The following parameters are assumed: the depth to the reflector, $D_r = 4000$ m; velocity in the overburden, $V_{dat} = 3000$ m/s; the spread length, $L = 2000$ m. The response to the left side of the anomaly alone is denoted by R_1, the response to the right side alone by R_2, and the actual response by R_s. Figure 15 shows MCS velocity variations for different widths of the velocity anomaly. Note that the response is, again, many times greater than the actual velocity anomaly and in the opposite sense to it. Note also that R_s is the result of interference between R_1 and R_2, the interference being, approximately, linearly additive,[29] i.e.

$$R_s \simeq R_1 + R_2 \tag{30}$$

This near-linearity is related to an observed near-linear relationship between the response and the magnitude of the delay (see also Levin[30]). R_s reaches maximum when the peaks of R_1 and R_2 become nearly coincident at an anomaly width of about $0.6 L$ (Fig. 15(D)).

In effect, the spread may be thought of as a filter of an impulse response, R_f. In this sense, the response curve becomes a 'convolution' of R_f with the statics (although this is not rigorously true). Changing the spread length is equivalent to a horizontal translation of the transform of R_f.

In view of the large variety of possible sources of statics (see preliminary remarks of this section), the spectrum of the statics can be expected to be fairly flat over ranges that are of interest in practice. It can thus be seen that MCS velocity oscillations will normally be present on the velocity profile,

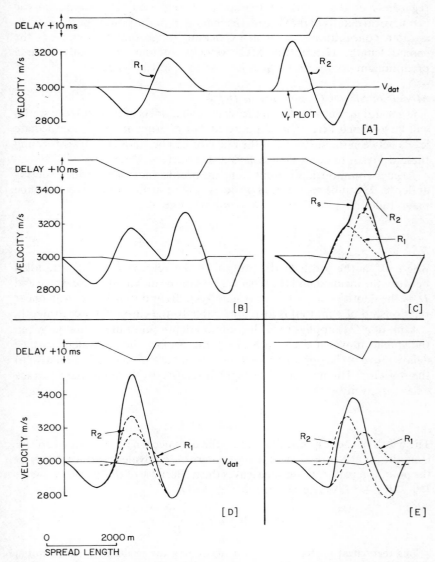

FIG. 15. MCS velocity profiles across shallow velocity anomalies (trough shape) of different widths.

regardless of the length of the spread being used. Depending on the geometrical details of the spread, the main part of the power spectrum of R_f is often concentrated about a wavelength range that encompasses the spread length. Hence, the MCS velocity oscillations usually have a predominant component which is close to a spread length.

Models of Statics Originating at Depth

The general model adopted in the section on models of near-surface statics will be assumed here. The response to delays originating at intermediate levels between the surface and the reflector can be studied by transforming these delays into equivalent delays at the surface.[29]

At each common depth point along the profile, the response due to delays at depth D_d can be reproduced by delays at the surface of exactly the same magnitude but spread over a horizontal range of

$$W_e = \frac{W_a D_r}{(D_r - D_d)} \tag{31}$$

where W_a is the width of the portion of the velocity anomaly straddled between the incident and the reflected rays of the maximum offset trace and D_r is the depth of the reflector (Fig. 16(A), (B) and (C)).† The near-linear relationship of eqn. (30) is also applicable to the case of statics at depth.

Equation (31) applies to each common depth point individually. When the spread moves a distance S along the profile, the equivalent surface delays move in the opposite direction, due to the 'parallax' effect relative to the reflector. The motion of the spread relative to the equivalent surface delays is given by

$$S_d = \frac{SD_r}{(D_r - D_d)}$$

The width of the response is given by the distance which the spread covers from the point where its leading edge enters the equivalent surface delays to the point where its trailing edge leaves them. Because of the 'parallax' effect, $W_R \neq W_e + L$. As can be seen in Fig. 16(D),

$$W_R = W_a + L - \frac{LD_d}{D_r}$$

† This theoretical model is useful for illustrating the problem. Fitch (editorial comment) points out that, in practice, the effect of wavefront 'healing' would reduce the delay. Thus, in Fig. 16(A), little or no delay might be observed; in Fig. 16(B) and 16(C), the full effect will be encountered in the central part of W_e, diminishing progressively towards the ends.

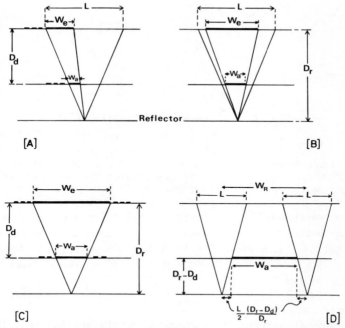

FIG. 16. (A)(B) and (C): Geometrical relationships between delays at depth and their surface equivalent for various positions of the spread relative to the delays. (D): The width of the response curve in relation to L, W_a, D_d and D_r.

Thus, although the width of the response is intuitively expected to increase with increasing delay depth, the 'parallax' effect produces the opposite result; the response curve is increasingly compressed as the ratio D_d/D_r increases. This compression is responsible for the frequent presence of wavelength components that are shorter than a spread length on velocity profiles.

Figure 17 compares the response curves resulting from placing the delay model of Fig. 15(D) at, (a) the surface and (b) a depth of 2000 m. Note that the compression of the response curve as D_d/D_r increases from 0 to 0·5 is accompanied by a reduction in amplitude. The amplitude reduction may be understood from a consideration of the equivalent surface delays; as the delay depth increases the width of the equivalent surface delays increases and their steepness (rate of change along the surface) decreases. Consequently, the distortion of the reflection curvature in the CDP gather diminishes. This phenomenon may also be viewed within the general

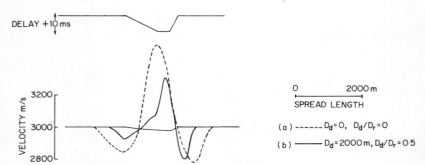

Fig. 17. Comparison of MCS velocity profile across a delay model when the delays are situated (a) at the surface and (b) at 2000 m depth. Note that increasing delay depth compresses the response curve in width and amplitude.

context of wavefront 'healing' mentioned in the footnote. It may thus be concluded that the response to statics originating at depth diminishes as the ratio of statics depth to reflector depth increases or, more specifically, near-surface statics have a greater effect on the velocity profile than those originating at depth.

Example of Field Profile

Figure 18 shows velocity profiles corresponding to three reflectors, A, B and C. Reflector A is at about 2·1 s (two-way travel-time), some 0·2 s above B, and 0·3 s above C. The interval between successive velocity analysis points along the line is 300 m. The maximum offset (spread length) is 2740 m. The dips on the reflectors are negligible.

The three profiles show many features that are commonly observed in practice. As shown on profile A, the measured MCS velocity varies with a short wavelength component superimposed on an oscillation of a wavelength close to a spread length. The short wavelength variations are due to random noise in the stacked data (see the section on noise). These variations have been smoothed out on the profiles.

The oscillations on the three profiles are generally 'parallel' owing to the fact that a given time delay on a trace is usually common to the three reflections. The general increase in the magnitude of the oscillations with increasing reflector depth is due to,

(a) the increasing relative effect of the delay as the NMO decreases with depth,
(b) the decrease in the ratio D_d/D_r with increasing reflector depth (see the section on models of statics originating at depth), and

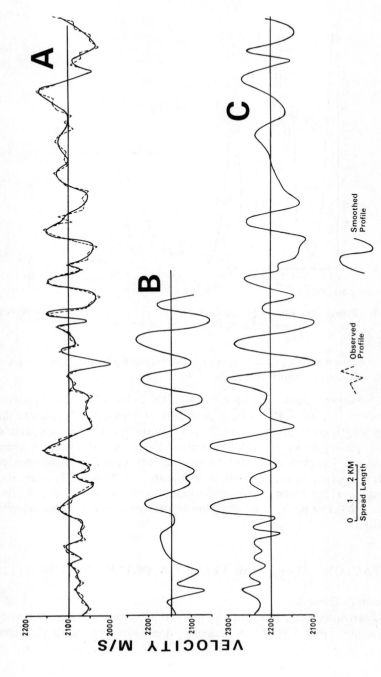

FIG. 18. MCS velocity plots corresponding to three reflectors, A, B and C along a seismic line.

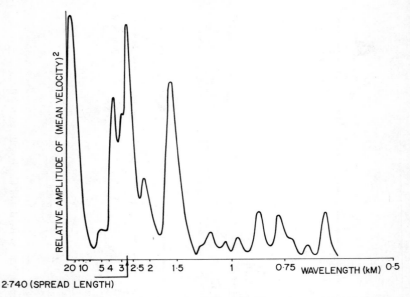

FIG. 19. Power spectrum of profile C of Fig. 18. Note the prominent peak at a wavelength equal to a spread length.

(c) the increasing probability of encountering further delays along the raypath as depth increases.

The power spectrum of profile C (without DC or linear trend component) is shown in Fig. 19.[31] The conspicuous peak at a wavelength equal to the spread length corresponds to oscillations caused by near-surface statics. The subsidiary peak at 3·5 km is within the expected range of these oscillations. The peak at about 1·5 km represents, mainly, oscillations due to delays originating at intermediate depths above reflector C. It could also be, partly, a harmonic of the main component (of about double this wavelength). The peak at about 20 km is attributed to a genuine velocity variation.

FACTORS AFFECTING VELOCITY DETERMINATIONS

Preliminary Remarks

It has been noted in the introduction and in the section on time and velocity relationships that the MCS velocity does not represent the true propagation velocity except in the ideal situation where the ground consists of a single horizontal reflector beneath a uniform overburden. In practice, wide deviations from this ideal model cause the determined MCS velocity to become a physically meaningless quantity, though still retaining the dimensions of velocity. Factors related to these deviations (e.g. refraction) do not affect the accuracy of the MCS velocity as a parameter. They enter into play at the MCS–r.m.s. velocity derivation stage (Fig. 2). Other factors (e.g. processing errors) are not related to the simplicity of the assumed model. They produce inaccuracies in the determined MCS velocity even over an ideal ground. The derivation of interval and average velocities involves fewer sources of error as indicated in Fig. 2.

The main factors affecting velocity determinations will now be reviewed. When possible, methods for determining the velocity error are indicated so that an appropriate correction may be made to reduce or remove the error. Otherwise, methods for estimating the accuracy of the derived velocity are given. In a few cases, only a qualitative discussion is possible.

The factors are reviewed according to the following classification:

(1) Acquisition errors;
(2) Processing errors;
(3) Noise;
(4) Errors related to wavelet form;
(5) Errors related to wave propagation;
(6) Velocity and structural variations in the ground;
(7) Subjective errors.

This classification is based on the scheme of Fig. 2 but does not follow it exactly. It is not intended to be a rigid classification. Some factors are common to more than one class while some classes could themselves be subdivided or included under others.

Acquisition Errors

Generally, marine data are more prone to acquisition errors than land data owing to the difficulty in controlling the acquisition parameters in the case of the former.

Offset Errors[32]

Constant error. This situation arises, for example, when the offset of the near trace is incorrectly estimated so that the same error is repeated in the estimate of the remaining offsets. For a constant offset error Δx, the error

ΔV in the estimated velocity, V_{MCS}, is given by

$$\frac{\Delta V}{V_{\text{MCS}}} \simeq \frac{\Delta x}{X_m} \tag{32}$$

where X_m is the maximum offset, assumed to be much greater than the near trace offset. An overestimated offset produces an overestimated velocity. A 12 m error in the offsets of a 2500 m cable produces a velocity error of 0·5 %.

Error proportional to offset. This situation is closely approximated, for example, when there is a significant streamer curvature. If the constant u represents the ratio of the assumed offset to the true offset then

$$\frac{\Delta V}{V_{\text{MCS}}} \simeq u - 1 \tag{33}$$

Irregular offset errors. If the errors are random, with a standard deviation, σ_x, then the standard error of the velocity estimate is given by

$$\sigma_v \simeq \frac{2 \cdot 3 V_{\text{MCS}} \sigma_x}{(X_m m^{1/2})} \tag{34}$$

where m is the number of traces. The value of σ_v increases slightly when the inner trace offset exceeds one group interval. For $V_{\text{MCS}} = 2500$ m/s, $\sigma_x = 5$ m, $X_m = 2500$ m, and $m = 24$, $\sigma_v = 2\cdot 4$ m/s. If the error in each offset is known then the resulting velocity error can be estimated approximately by substituting the true and the assumed offsets in eqn. (11) and comparing the results.

Ship Motion

The distance travelled by the ship between the moment of explosion and the recording of each reflection introduces a velocity error (overestimation) which is equivalent to a small offset error. For a given reflection, the error consists of a constant offset error component, which may be estimated from eqn. (32), and a negligible component proportional to offset.

Streamer Feathering

Consider a single plane reflector with maximum dip ϕ at an angle θ with the profile. The travel-time, T_x, for a trace offset X in a straight streamer at a feathering angle, f, is given by

$$T_x^2 = \frac{4d^2 + X^2 - 4dX \sin \phi \cos(\theta - f)}{V^2} \tag{35}$$

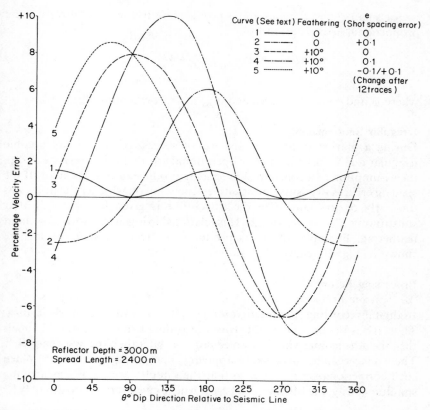

FIG. 20. Velocity error caused by feathering and shot spacing error in the presence of dip. After Lucas.[33]

where d is the perpendicular distance between the source and the reflector and V is the velocity in the overburden.[33] θ and f are measured in the same sense. The perpendicular distance between the common ground point (along the profile) and the reflector, D, is related to d by

$$D = d - 0.5X \sin \phi \cos \theta \qquad (36)$$

Generally, for small feathering angles, the velocity error increases from a minimum for profiles close to the dip direction to a maximum for profiles close to the strike.[34] Figure 20 shows an example of the percentage velocity error as a function of θ for the feathering-free case (curve 1) and a $+10°$ feathering (curve 3), ϕ being $10°$.

For small ϕ and f angles and a profile close to the strike direction the proportional error is given by

$$\frac{\Delta V}{V} \simeq 0.0006 \frac{D\phi f}{X_{max}}$$

where ϕ and f are in degrees and X_{max} is the maximum offset.[34]

Irregular Shot Spacing[33]
During a marine survey, the spacing of successive 'pops' may become irregular due to instrumental or operational errors. In the presence of dip, the assumption of a constant shot spacing produces a velocity error. If the spacing error is e (positive or negative depending on whether the firing was too early or too late) then the travel-time is given by eqn. (35) after substituting $X'(=X+e)$ for X. This relationship is general, with or without feathering. Examples of errors produced by irregular shot spacing are shown in Fig. 20 (curves 2, 4 and 5).

Processing Errors
Multiplexer Error
In digital recording, the multiplexer samples successive channels in turn. Thus, if p is the time interval between recording two consecutive channels then the qth channel will be sampled $p(q-1)$ later than the first channel.[35] These delays can be corrected by applying appropriate time shifts.[36] When such corrections are ignored, so that the samples are treated as arriving simultaneously, the MCS velocity error will be given very approximately by

$$\Delta V \simeq -\frac{p V_{MCS}^3 T_0}{m y^2} \qquad (37)$$

where m is the number of stacked traces and y is the channel spacing.[34] If channel 1 corresponds to the far trace, as is usually the case, the velocity will be underestimated. Typically the error is about $0.3-1.5\%$.[34]

General Pre-velocity Analysis Process[34]
Whitening. When the fold of cover is high, the addition of random noise to the data does not have a significant effect on the reliability of the velocity analysis. The general subject of random noise is discussed later.

Band-pass filtering. Generally, short pulses and high predominant frequency enhance the resolution on the velocity analysis display.

Data normalisation. Comparison of results obtained from automatic and programmed gain control suggests that better velocity analysis results are obtained when the former method is employed.

Datum corrections. The proportional velocity error resulting from a constant time shift applied to the entire gather may be computed from eqn. (40), as shown later. Note that this error decreases with increasing time.

Common offset stacking. See the section on offset-related wavelet changes.

Details of the Analysis Process and Output
The accuracy of the MCS velocity value picked from the velocity analysis display is influenced by the resolution achieved by the analysis algorithm. The principal algorithm parameters which influence the resolution are;

(a) the length of the analysis gate which, to obtain fine resolution, should not exceed the predominant period of the reflection wavelet,
(b) the interval between successive gates in the computation and output, and
(c) the step by which successive velocity (or NMO) values are incremented, in the computation and output.

The velocity accuracy is also somewhat influenced by the details of the coherency criterion and by the weight that might be attached to various traces.

Noise
Coherent Noise
Coherent ambient noise (e.g. that due to cable snatch at sea) is horizontally travelling. Therefore, it has very low velocity across the spread and would be readily distinguishable on the velocity analysis output.[34] Non-hyperbolic coherent events are strongly discriminated against by the velocity analysis process and need not concern us here. Other coherent events which are not part of a primary reflection (e.g. diffractions) are discussed under relevant headings.

Random Noise
For the present purpose, included under the term random noise are such effects as instrumental noise, natural noise, incoherent seismic interference, imperfect statics corrections (of much shorter wavelength than a spread

length—cf. the section on lateral resolution and statics), etc. In velocity analysis work, the effect of random noise is equivalent to, and may be simulated by, random time shifts (or jitter) superimposed on the noise-free travel-time curve. It may be assumed that these time shifts are normally distributed.[37]

In the presence of random time shifts of variance σ_t^2 the MCS velocity variance is given by

$$\sigma_v^2 = \frac{\sigma_t^2 V_{MCS}^2 \left(\sum_{i=1}^{m} T_i^2 x_i^2 \right)}{\left(\sum_{i=1}^{m} T_i^2 x_i \right)^2} \tag{38}$$

where

$$x_i = m X_i^2 - \sum_{j=1}^{m} X_j^2$$

X_i and T_i are respectively the offset and travel-time for the ith trace and m is the number of stacked traces.[14] An approximate but more convenient expression is

$$\sigma_v \simeq \frac{1 \cdot 6 \, \sigma_t V_{MCS}}{(m^{1/2} \Delta T_{max})} \tag{39}$$

where ΔT_{max} is the NMO of the outermost trace in the stack.[14,32]

Equation (39) shows that the noise-generated error is at its lowest when the NMO reaches its maximum and the full fold of cover is used. Such conditions are usually attained at about 1·5–2·0 s, below which the error increases as the NMO diminishes and the value of V_{MCS} increases.[38] Equations (38) and (39) are based on simplifying assumptions which ignore the non-hyperbolic shape of the reflection in the CDP record. Because of this simplification, the actual improvement in the signal–noise ratio is, invariably, smaller than $m^{1/2}$.[36]

The histogram of Fig. 21 illustrates the statistical distribution of MCS velocity error in the presence of random noise. The histogram is based on 300 MCS velocities corresponding to the same reflector but each MCS velocity is obtained with a different set of random time shifts of a standard deviation of 4·6 ms. The relevant model parameters are listed on the figure.

The velocities of the histogram are normally distributed. The slight skewness towards high velocities is equivalent to the anti-symmetry observed in Fig. 13. Note that the mean is very close to the zero error

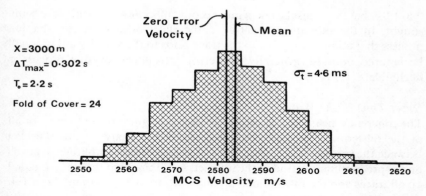

FIG. 21. Histogram of MCS velocities in the presence of random noise. The velocities correspond to a horizontal reflector in a simplified model similar to that of Fig. 6. Note that the mean falls close to the zero error value.

velocity which is the MCS velocity value that would have been obtained had the data been noise-free.

Practical observations conform well with the foregoing considerations. MCS velocities measured along a closely sampled profile fluctuate with a standard deviation close to that predicted by eqns. (38) and (39). Profile A, Fig. 18, shows an example of such fluctuations.

Generally, a smooth mean curve through the fluctuations should correspond closely to the MCS velocity profile that would be measured in the hypothetical absence of random noise, on the basis that the mean of the MCS velocity distribution falls close to the zero error value. For the same reason, the maximum coherency build-up on the velocity analysis display is usually quite stable in the presence of random noise. The stability is enhanced by the fact that the derived velocity represents an average, not only across the gather but also along the analysis gate. Thus, reliable velocity estimates are usually obtainable even when the noise level is several times that of the reflection.[34]

Errors Related to Wavelet Form
General Remarks
From the discussion of velocity measurement techniques (see the section on current techniques) it is clear that the actual shape of the reflection wavelet is fundamental in determining the accuracy and resolution of the velocity analysis process. Individual wavelet-related errors do not usually exceed

0·5–1·0% but they must be taken into account if accurate velocities are being sought. In the subsequent treatment, the effect of any wavelet shaping process that might be applied to the trace prior to the velocity analysis will be ignored because, generally, the error introduced by this process is negligible.

Onset Time of the Wavelet

The coherency measure does not attain its maximum value until most or all of the reflection wavelet is contained within the analysis gate. The lag between the true onset of the wavelet and the maximum amplitude depends on the predominant period of the wavelet. Assuming that the lag, t, is equal on all traces and the ratio of maximum offset/depth is less than about 1·4, which is the usual case in practice, then[15]

$$\frac{\Delta V}{V_{\text{MCS}}} \simeq \frac{t}{2T_0} \tag{40}$$

Thus, for a lead of 15 ms at zero offset time of 1·5 s the MCS velocity will be underestimated by about 0·5%. If it can be assumed that the lag is proportional to travel-time then[39]

$$\frac{\Delta V}{V_{\text{MCS}}} \simeq -\frac{t}{T_0} \tag{41}$$

Offset-Related Wavelet Changes

The factors affecting wavelet shape produce distortions that mostly increase with increasing offset. The main consequence is a smearing of resolution on the velocity analysis output. Further loss in accuracy results from possible change in the lag between the wavelet onset and the maximum coherency build-up. If the lag can be treated as being proportional to travel-time then eqn. (41) would apply. The main factors producing offset-related wavelet changes are as follows.

Absorption effects of the ground. These increase with increasing frequency, so that the ground acts as a low pass filter. As the offset increases, the wavelet travels longer in the ground and hence becomes differentially richer in low frequencies.

Thin sedimentary layering. This produces phase and amplitude distortions which increase with increasing offset.[40] This effect increases the dependence of the velocity estimate on the bandpass filtering used because the thin-layer interference phenomenon is frequency dependent.[38,41]

The response of a summed linear array of geophones. A reflected plane wave is not received simultaneously by individual geophones within an array. The time delay between the first and last geophone in an array of length r is given by

$$t = \frac{r \sin E_0}{V_0} \qquad (42)$$

where E_0 is the angle at which the plane wave reaches the surface and V_0 is the velocity in the top layer. Therefore, at large offsets (large E_0), the summed response can distort the wavelet significantly.

Figure 22 illustrates the change in the wavelet shape as it is received by arrays at successively larger offsets.[15] There is an increase in the wavelet duration and a decrease in its amplitude. These changes are generally accompanied by deterioration in the signal–noise ratio on the far traces. The deterioration reaches a maximum when t becomes equal to the predominant period of the reflection wavelet.[34]

Common offset (vertical) stacking. This produces wavelet distortions when there are lateral variations in the section such as rapid character changes or significant dip. In the presence of dip, ϕ, the change in the zero offset time over horizontal distance H is[34]

$$t \simeq \frac{2H \sin \phi}{V_{\text{MCS}}} \qquad (43)$$

The wavelet distortions resulting from vertical stacking are analogous to those produced by a summed array as indicated by the similarity between eqns. (42) and (43). However, in the case of vertical stacking, the wavelet distortions do not increase with increasing offset. For large spread length–depth ratio, the distortion decreases at large offsets.

Levin[42] has considered the filtering effect of vertical stacking in the presence of dip. The following model is used as an illustration:

$V = 2500 \, \text{m/s}$, $H = 40 \, \text{m}$, $\phi = 4°$ (predominant frequency 40 Hz), N = number of summed traces, A = ratio of (vertically) stacked amplitude to unstacked amplitude, D = reflector depth, and X = offset. We have:

At $X = 0$ (X–D ratio = 0); for $N = 2$, $A = 0.95$, and for $N = 6$, $A = 0.60$.
At $X = 1000 \, \text{m}$, $D = 3000 \, \text{m}$ (small X–D ratio); similar to the above.
At $X = 3000 \, \text{m}$, $D = 1000 \, \text{m}$ (large X–D ratio); for $N = 2$, $A = 0.98$, and for $N = 6$, $A = 0.85$.

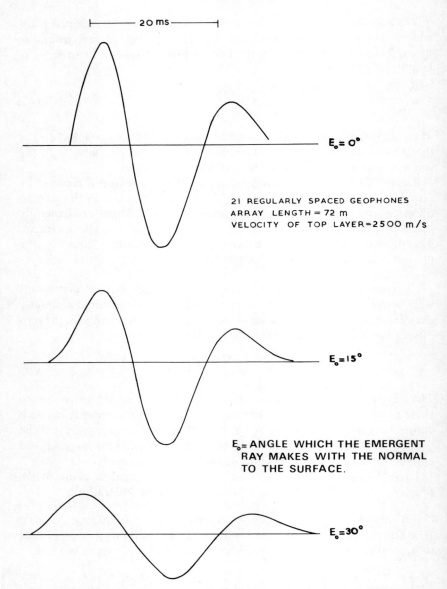

FIG. 22. Variation of the response of a summed linear array of geophones with E_0. The pulse duration increases and its amplitude decreases with increasing offset (increasing E_0).

There is a corresponding reduction in the predominant frequency of the output trace as N increases.

NMO corrections. These cause wavelet stretching that increases with offset. Some degree of asymmetry is introduced in the stretched wavelet but the amplitude remains unchanged. The stretching of the wavelet compresses its spectrum. The stacked pulse is richer in low frequencies than is anticipated and the signal–noise enhancement is smaller than would be obtained in the absence of stretching.[43]

Errors Related to Wave Propagation
Refraction

It was indicated in the section on MCS and r.m.s. velocities, that increasing refraction along the raypath trajectories corresponding to the traces of the CDP gather increases the bias in the estimate of the r.m.s. velocity from the MCS velocity. In many practical applications, the bias is sufficiently small as to be negligible. However, a large bias caused by severe velocity variations and/or through the use of a long cable should be corrected. For example, if MCS velocities are used as r.m.s. velocities in eqn. (20) to determine the velocity of the seventh layer in the model of Fig. 6, the result will be too large by 10%, as predicted by eqn. (48). The bias can be removed in a number of ways as follows.

The shifting stack technique.[14] This is based on determining successive MCS velocities from different groupings or traces in the CDP gather. The effective offset (defined in eqn. (14)) of each group is plotted vs the corresponding MCS velocity. The extrapolated MCS velocity value at zero effective offset gives the required r.m.s. velocity. The technique is quite useful, if cost can be justified, because it serves at the same time to correct for other factors which vary systematically and semi-systematically with offset. The determination of V_{NMO} (see the section on dipping reflectors) is one such application.

Stacking according to a three-term truncation of eqn. (2). This suggests itself as a possible method for a bias-free determination of $V_{r.m.s.}$ in view of the accurate time–distance relationship which this truncation produces (see the section on convergence properties of eqn. (2)). However, statistical considerations indicate that, in the presence of random noise, the standard deviation of the determined velocity is often too large for practical purposes.[14]

Estimation from eqn. (13). This equation provides a rapid method for obtaining an estimate of the bias with a precision that is adequate for many purposes. The measured MCS velocity may be substituted for $V_{r.m.s.}$. Gross interval velocities derived through eqn. (20) can be used for the estimate of v_k. X_e is usually about 0·6 of the maximum offset.

Anisotropy
It has long been observed that stratified rocks are frequently anisotropic, i.e. the velocity along the direction of bedding is different from (almost invariably higher than) the velocity across it.[44] Anisotropy can therefore produce a large effect on the measured MCS velocity. Variation of velocity along the bedding plane (transverse anisotropy) rarely reaches a significant level. It will be assumed that the layers are transversely isotropic.

In its broad sense, anisotropy can take the following forms:

(1) Micro-anisotropy, which is an intrinsic property of the anisotropic medium. This is the form usually implied in common usage.
(2) Macro-anisotropy. A medium consisting of a series of parallel isotropic homogeneous layers responds on average as an intrinsically anisotropic medium when the predominant wavelength greatly exceeds the layer thickness.[45]
(3) Quasi-anisotropy, associated with refraction at layer interfaces.[46]

There is a degree of gradation between the three forms. The ratio of velocity along the bedding to that normal to the bedding may be expressed by the parameter A'. Values of A' ranging from 1·0 to 1·4 have been quoted in the literature.[46]

Anisotropy distorts the wavefront from the semi-circular shape of fully isotropic propagation. In general, the resulting wavefront is not an ellipsoid. However, for angles of incidence not exceeding 30° the incident part of the wavefront can be approximated by a portion of a 'best fit' ellipsoid.[47] Three elastic constants are sufficient to describe the wavefront in this region. The anisotropy factor, A, is defined as the ratio of the horizontal axis to the vertical axis of this ellipsoid.[48]

For small angles of incidence and a horizontal reflector,

$$T_x^2 = \left(\frac{D}{V}\right)^2 + \left(\frac{X}{AV}\right)^2 \qquad (44)$$

where D is the depth to the reflector and V is the propagation velocity in the vertical direction. The MCS velocity is an estimate of AV.

Note that the velocity along the bedding is $A'V$, not AV. An ellipsoid whose horizontal–vertical axis ratio is equal to A' generally bears little relationship to the shape of the propagating wavefront. Hence, A' and A are usually quite different. In fact, as a result of the particular detail of the wave field in the region of incidence, it is possible that the vertical axis of the best fitting ellipsoid will be greater than the horizontal axis. Consequently, A will be less than unity. In such cases, the MCS velocity value could become smaller than the average vertical velocity even though A' might be greater than unity.[47] Clearly then, the ratio of the velocity along the bedding to that across the bedding cannot be reliably used to estimate the anisotropy factor.

In some plausible anisotropic earth models, errors of up to 20% could result in time-to-depth conversion if the MCS velocities were not adjusted to the vertical average velocities.[47] The value of A could be estimated when estimates of the relevant elastic constants are available. The shifting stack technique (see the section on refraction) is suggested as another possible way of dealing with anisotropy. It provides an estimate of the r.m.s. velocity along the normal incidence raypath. When only gentle dips are involved, such a raypath is not influenced significantly by the anisotropic effects of the ground.

Multiples

Horizontal layering. The time–distance relationships for primary reflections (see the section on the time–distance relationship) apply equally to multiples provided that the assumed conditions of uniform horizontal layering are not violated significantly. A coherency build-up associated with a partly or wholly multiple trajectory will result when a stacking velocity appropriate to that trajectory is applied. This velocity is a first approximation to the r.m.s. velocity defined in eqn. (45).

Along any trajectory some of the layers are traversed down and up once only (primary part) while others are traversed more than once (multiple part). The r.m.s. velocity along the trajectory may be re-defined from eqn. (3) as

$$V_{\text{r.m.s.}}^2 = \frac{\left(\sum_{k=1}^{p} v_k^2 t_k + \sum_{j=1}^{m} v_j^2 t_j\right)}{T_0} \tag{45}$$

where p and m are the number of primary and multiple traverses respectively. Equation (45) shows that the r.m.s. velocity value will be biased towards the velocities of those layers traversed by the multiple part of the raypath.

The trajectory of a multiple is contained within a shallower part of the ground than a primary of a similar zero offset time. Because of the usual increase of velocity with depth, multiple reflections generally correspond to a lower velocity than primaries of the same zero offset time. However, multiples corresponding to higher velocities than primaries of similar zero offset time are also encountered in practice. For example, in the model of Fig. 6, a multiple which is reflected within the seventh layer will correspond to a higher velocity than a primary reflected from a deeper interface, in accordance with eqn. (45).

The difference between the primary velocity and the multiple velocity produces differences in coherency build-ups on the velocity spectra and in the reflection curvature in the CDP gather; the smaller the velocity the larger the curvature. These differences are useful features for distinguishing and isolating multiples in velocity work (see the section on general features and interpretation of velocity spectra displays). Often, however, there is interference between the multiple and the primary reflections, causing various distortions on the velocity analysis display. These distortions can impede the determination of reliable MCS velocity values.

Dipping interfaces. The presence of dip introduces many complications. The simplest ground model is a dipping reflector below a uniform overburden of velocity V. In this case, the velocity corresponding to the maximum coherency is given by

$$V_m = \frac{V}{(1 - \cos^2 \theta \sin^2 (M + 1) \phi)^{1/2}}$$

where ϕ is the true dip, θ is the angle between the direction of the true dip and the profile, and M is the order of the multiple, being zero for a primary, one for a three-bounce reflection, etc.[18] Thus the measured velocity value increases with the order of the multiple, being equivalent to the MCS velocity for a primary from a reflection whose dip is $(M + 1)\phi$.

In practice, where the ground consists of a large number of layers with varying dips and where both peg legs and simple multiples exist, the problem can become quite intricate. Different peg leg multiples may follow different trajectories but arrive at the same time at zero offset. These multiples generally have different travel-times at offsets other than zero and therefore correspond to different velocities. A common occurrence of such peg legs is in association with water reverberations in marine surveys.[49] Levin and Shah[50] give a useful account of peg legs in the presence of dipping reflectors.

Diffractions

Diffractions are a possible source for unexpectedly high velocities sometimes encountered in practice.[51] For a distant diffractor in a uniform ground, the velocity producing maximum coherency is related to the propagation velocity, V, by

$$V_{sd} = \frac{V}{\cos \alpha} \qquad (46)$$

where α is defined as follows; the straight ray from the common ground point (CGP) to the diffractor is projected onto the vertical plane containing the seismic line—α is the angle between this projection and the normal to the surface.[51]

It will be recognised that eqn. (46) is similar to eqn. (23) for a dipping reflector (see the section on dipping reflectors). The correspondence between diffractors and dipping planes can be understood by treating the diffractor as a distant spherical reflector. In this way, α will be the dip, in the direction of the profile, of the plane tangent to the sphere (i.e. the local dip) at the point where the ray from the CGP to the centre of the sphere intersects the sphere.

Coherency build-ups produced by diffractions can be differentiated from those produced by primaries by their rapid progressive change when followed on successive closely spaced velocity spectra. Point diffractors outside the plane of the profile and line diffractors at small angles to the profile produce a slower change but they are recognisable on cross profiles. Difficulties could arise in situations involving, for example, isolated velocity analyses near a line diffractor. Generally, however, coherency build-ups due to diffractions do not escape detection; they are at least identified as spurious maxima, if not specifically as diffraction-generated.

Shear Wave Propagation[34]

When a seismic wave is reflected at a boundary there is generally some conversion from compressional (P) waves to shear (S) waves and vice versa. The amount of conversion, which is zero at normal incidence, increases with increasing incidence angle. A great deal of shear wave energy is generated when using surface sources on land. Therefore, it is expected that some of the events which correspond to low velocity on the velocity analysis display represent not multiples but primaries that have travelled part of the path as shear waves.

In cases involving, for example, large dips, rapid velocity increases and/or large offsets, the critical angle at some interfaces may be approached or even exceeded. Under these conditions, significant phase, amplitude and mode changes take place.[52] These changes can lead to serious velocity errors if the associated events are not discriminated. However, reliable velocity work is rarely carried out in these circumstances. P and S wave conversion is not a factor that needs to be reckoned with unless very accurate velocities are being sought.

Velocity and Structural Variations in the Ground
Laterally Variable Time Delays (Statics)
It has been noted in the section on lateral resolution and statics, that variable time delays above the reflector produce oscillations in the MCS velocity profile. The magnitude and width of these oscillations vary from one area to another according to the details of the ground above the reflector. Typically, the oscillations (peak to trough) are about 5–10% of the true average velocity but oscillations reaching 30% or more are not uncommon.[53] Such oscillations are generally the largest sources of error in velocity determinations. They emphasise the importance of producing velocity analyses at appropriately close intervals; isolated velocity measurements can produce misleading results. The velocity profile should preferably be several times the length of the spread, avoiding as far as possible, interruptions by faults and other abrupt structural changes.

There is, as yet, no standard way of correcting the velocity profile for these oscillations. A number of possible approaches will be reviewed.

Filtering. Filtering the oscillations out of the velocity profile by averaging the profile over one or two spread lengths[38] is an obvious extension of the fact that the oscillations are predominantly of a wavelength which is close to a spread length. Figure 23 shows the results of applying one-spread length and two-spread length operators to the velocity profile corresponding to reflector C of Fig. 18. The effectiveness of the filtering is quite clear. In this particular example, the finally smoothed velocity profile is based on intermediate values between the two filtered versions.

The filtering process reduces the lateral resolution along the profile and, as shown by Fig. 23, does not completely remove the oscillations. Also, the averaged values are asymmetrical with respect to the true average, having a slight bias towards higher velocities (see the section on the step model). However, these are relatively minor limitations. The filtering approach is a practical way of treating the problem of velocity oscillations.

In the example of Fig. 23, the averaging was based on the mean value. Fitch (editorial comment) suggests that the median is more robust than the

VELOCITY DETERMINATION FROM SEISMIC REFLECTION DATA

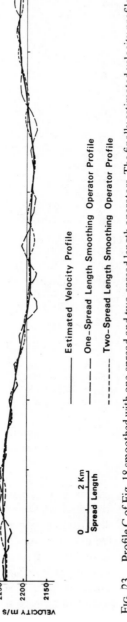

Fig. 23. Profile C of Fig. 18 smoothed with one-spread and two-spread length operators. The finally estimated velocity profile is based on values between the two smoothed versions.

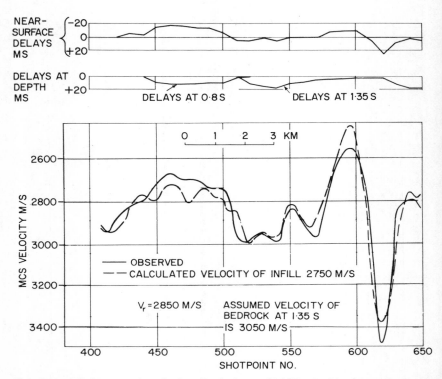

FIG. 24. A fit between the calculated and observed MCS velocities for a reflector at an average depth of 3000 m. The delays are assumed to originate at the surface and at two levels at depth, the deep delays being postulated from an interpretation of the seismic section.

mean for averaging purposes yielding generally stable, slowly changing values. The median also avoids the slight bias towards higher velocities (see the section on the step model).

Modelling techniques. An estimate of the delays which generate the oscillations may be obtained through modelling techniques. A model of the ground is constructed from all available information. A set of assumed delays is then iteratively adjusted until the calculated MCS velocities fit the observed profile.[54] The distribution of delays need not be restricted to one level. Figure 24 shows an example of delays computed in this way and the resulting fit between the calculated and the observed MCS velocities.[53]

The solution by this method is non-unique. A large number of time delay sets can be found which, when suitably combined with velocity variations, produce a good fit on the velocity profile. Therefore, this approach is more useful for confirming an assumed velocity distribution for which there is already some evidence.

Finding a set of delays. An approach related to the preceding one consists of finding a set of delays which when removed from the appropriate traces would enhance the coherency in all of the CDP gathers along the profile.[55] The dynamic and static corrections are, thus, locked up in a loop that continually improves the estimates of both. In order to keep the treatment simple, the delays are usually assumed to originate close to the surface.

'*Deconvolving*'. The near-linear relationship between the delays and the response curve (see the section on the trough model) leads to the possibility of 'deconvolving' the velocity profile by the response to a spike delay to obtain an estimate of the delays causing the oscillations. Potentially, this approach is quite powerful. Difficulties arise in practice from the presence of 'noise' in the input velocity profile.[56]

Plane Dipping Reflectors
In the presence of arbitrarily dipping plane reflectors, the interval and average velocities along the normal incidence raypath can be derived directly from the normal moveout velocity, V_{NMO} (see the section on dipping reflectors). The accuracy with which these velocities are obtained varies widely according to the validity of the assumed model, viz. constant velocity within each layer, constant reflector dip, absence of pinchouts, etc. An error in the estimated interval velocity of one layer produces errors in the parameter estimates of all deeper layers.

V_{NMO} may be estimated from MCS velocities by using the shifting stack technique (see the section on refraction); the extrapolated MCS velocity value at zero offset (normal incidence) gives an estimate of V_{NMO}. The technique is applicable equally to two-dimensional and three-dimensional problems. Errors arising from streamer feathering, vertical stacking, multiples, etc., in the presence of reflector dip are discussed under relevant headings.

Complex Structures
In order that the hyperbolic time–distance relationship should hold sufficiently approximately, both legs of the reflection path should encounter

every layer above the reflector with the same dip and velocity.[57] Ground conditions seldom satisfy this requirement exactly.

Slight reflector curvature can be ignored for most practical purposes. Krey[20] deals with the general problem of curved reflectors. When the reflectors are significantly non-linear, the derivation of subsurface velocities from MCS velocities does not lend itself to simple analytical treatment. Iteratively adjusted ray tracing modelling can be a useful alternative.

In areas of severe tectonism or complex velocity distribution, even the ray tracing approach becomes prohibitively difficult to implement. In such cases, no reliable velocity determinations are possible.

Velocity Heterogeneity within Individual Intervals
An interval velocity computed from eqn. (20) overestimates the average velocity in the interval, v_a, by an amount that varies according to the velocity heterogeneity within the interval (see the section on interval velocity). From eqn. (17), the overestimation is given by

$$\Delta \bar{v} \simeq 0.5 \, v_a g \tag{47}$$

In cases where the interval is considered to be sufficiently uniform, the correction $\Delta \bar{v}$ will be unnecessary. The error in the estimate of $\Delta \bar{v}$ depends on the details of its derivation (see the sections on average velocity and interval velocity). The resultant error in v_a is proportionately smaller.

Subjective Errors

The interpretation of velocity analysis output is subject to the interpreter's experience in identifying relevant events and isolating non-primary interferences. It is also subject to the accuracy in picking velocity and time values, which is largely determined by the resolution on the velocity analysis output. Factors affecting the resolution are discussed under relevant headings, in terms of frequency and pulse shape, random and coherent interferences, algorithm parameters, etc.

The subjective element in the derivation of interval and average velocities arises in connection with the identification, selection and timing of relevant horizons on the seismic section and from its correlation with the velocity analysis output. The error resulting from the subjective element in velocity determinations cannot be readily quantified because of the nature of the parameters on which it depends. Generally, however, the error is expected to be within 0·1–3%, providing that there is no significant misinterpretation of the data involved.

ESTIMATION OF INTERVAL VELOCITY ERROR

General Remarks
From our discussion of factors affecting velocity determinations, it is clear that the derived r.m.s. velocity is an imprecise estimate. Tight control at various stages between the acquisition of the data and the derivation of the r.m.s. velocity could significantly avoid, account for, or remove some of the errors. However, there are always errors that cannot be treated in this manner. In other cases, control may be hampered by practical conditions. All residual r.m.s. velocity errors are passed into the estimates of the interval and average velocities.

Errors of Known Sense
Let us consider the case where the interval velocity, v_{int}, is being determined in eqn. (20). Suppose that e_a and e_b are errors in V_a and V_b, respectively, and that each error is known to be an overestimation (positive) or underestimation (negative). Such situations arise when, for example, the bias or the effect of dipping reflectors are uncorrected. The interval velocity error is then

$$E \simeq \frac{(e_b V_b T_b - e_a V_a T_a)}{(T_b - T_a) v_{int}} \tag{48}$$

i.e.

$$E \simeq \frac{(D_b e_b - D_a e_a)}{h} \tag{49}$$

where D_a and D_b are the depths to the top and bottom of the interval and h is the thickness of the interval.

Some individual components of e_a and e_b are produced by factors that are common to the top and bottom of the interval (e.g. refraction, offset errors, statics, etc.). These components will be in the same sense (positive or negative) and, therefore, will tend to cancel each other out in eqns. (48) and (49). In fact, when

$$e_a \simeq e_b \simeq e$$

then we get from eqn. (49)

$$E \simeq \frac{e(D_b - D_a)}{h} \simeq e$$

This is a relatively small interval velocity error.

Precision Errors

If f is a function of m parameters, $p_1, p_2, \ldots, p_m$, which are known to a precision $\alpha_1, \alpha_2, \ldots, \alpha_a$, respectively, then the error α_f in f may be estimated from

$$\alpha_f^2 = \left(\frac{\partial f}{\partial p_1}\right)^2 \alpha_1^2 + \left(\frac{\partial f}{\partial p_2}\right)^2 \alpha_2^2 + \cdots + \left(\frac{\partial f}{\partial p_m}\right)^2 \alpha_m^2 \tag{50}$$

$$\sigma_{int} = \frac{(\sigma_a^2 V_a^2 T_a^2 + \sigma_b^2 V_b^2 T_b^2)^{1/2}}{(T_b - T_a)v_{int}} \tag{51}$$

where σ_{int}, σ_a and σ_b are the standard errors in v_{int}, V_a and V_b, respectively. Note that, in this case, the errors are additive.
If

$$\sigma_a \simeq \sigma_b \simeq \sigma$$

then, from eqn. (51),

$$\sigma_{int} \simeq \frac{1 \cdot 4 \sigma D}{h} \tag{52}$$

where D is the average depth of the interval. Equation (52) and, indirectly, eqn. (49) and (51), show that the interval velocity error is proportional to the interval depth–thickness ratio. When this ratio is very small (e.g. 20 m interval at 1000 m depth) the magnification of error is so great that no reliable interval velocity can be obtained. In practice, the selected interval thickness is usually a compromise between the need for a thin interval to maximise the resolution and a thick interval to minimise the error. Note that an interval velocity estimate obtained from eqn. (20) is independent of errors in the velocity estimates of other intervals.

Equation (50) also provides a useful way for estimating the error in cases involving, for example, interval velocity determinations by ray tracing or average velocity determinations from eqn. (16).

GENERAL USES OF VELOCITIES

Preliminary Remarks

CDP-derived velocities have a wide range of applications. Some of these applications are specific to certain problems or areas, others are more general. Only those applications that are currently the most common will be considered. These applications are listed in Table 1.

Clearly, in order that the investigated velocity variation should be detectable, its magnitude must stand well above the level of uncertainty (or error) in the determined velocity value. Precision requirements, therefore, vary widely from one problem to another.

Uses of Velocities in Seismic Processing
Stacking
The original and most direct use of stacking velocities is for producing CDP stacks. The applied stacking velocity need not be that producing maximum coherency. For example, a higher velocity than the MCS velocity is frequently used to enhance multiple suppression. In other instances, a lower velocity improves the quality of the stacked section. The use of stacking velocities which differ from the MCS velocity by up to 15% is not uncommon.

Migration
Correct migration of seismic sections requires adequate knowledge of velocities in the ground whether the migration is based on the Kirchoff integral or the finite difference approach. For a ground model consisting of a single dipping reflector, the migration velocity, V_m, along the dip direction is equal to the true overburden velocity, V. French[58] shows that, for a profile at an angle of θ to the true dip direction, the correct migration velocity is

$$V_m = \frac{V}{\cos \theta} \quad (53)$$

Thus, depending on the direction of the profile relative to the structure, the appropriate migration velocity could be less than, equal to, or greater than the MCS velocity. In the case of velocity varying with depth, V in eqn. (53) should be replaced by $V_{\text{Vr.m.s.}}$, the vertical r.m.s. velocity function which defines the moveout on a diffraction curve about a hypothetical scatterer.[58]

These considerations show that in cases involving strike variations there is usually no single migration velocity that would migrate all events correctly. In such cases, and when there are significant lateral velocity variations, a suite of migration velocity functions are usually attempted until an optimum migration is achieved.

Seismic Interpretation
The recognition of the effect of gross velocity variations is fundamental in

seismic interpretation.[36,59] 'Pull-ups' and apparent convergence of reflections at depth are examples of 'velocity' phenomena frequently observed on the seismic section. The identification of multiples is often aided by reference to velocity analysis displays (see the section on multiples). On the basis of well information, the age and/or predominant lithology of an interval may sometimes be recognised from its velocity. Such information can prove valuable in cases involving unconformities, correlation difficulties, and other interpretational problems. Some of the velocity-related interpretational problems are specific to the particular survey area, often requiring ray tracing modelling. The applications presented in the subsequent sections may also be regarded as special processes of interpretation.

Lithological and Stratigraphical Studies
General Remarks
    Rock velocities. Ranges of velocities of common rock types are shown in Fig. 25. The considerable overlap between these velocities rules out the possibility of identifying the predominant lithology of an interval purely

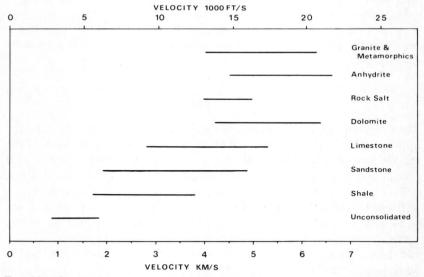

FIG. 25. Comparison between velocities of various rock types. The plot shows only the kind of ranges that are commonly encountered in practice. The full range is much wider.[2] Velocities corresponding to abnormal pressure conditions are excluded.

from its velocity. Frequently, the use of interval velocities is restricted to eliminating certain age or lithological possibilities.

Vertical and lateral differentiation. Generally, interval velocity resolution is easier to achieve vertically than laterally. This is because vertical lithological changes are normally more abrupt than lateral changes. Also, many of the factors affecting velocity determinations vary laterally, yet often produce approximately the same error at successive layer interfaces. These errors tend to cancel each other out (see the section on errors of known sense).

Accuracy Considerations
With well control in the area, the range of lithological possibilities is narrowed down to the extent that gross or even detailed lithological studies might become possible. However, the high precision requirements, particularly in detailed studies, can rarely be met by CDP-derived velocities. Consequently, the use of these velocities in lithological studies should be, primarily, to provide a general back-up to velocities derived from extensive well measurements. For example, suppose that correlations of a seismic interval to wells in an area reveal changes from carbonate facies of 4000 m/s velocity to shale facies of 2800 m/s velocity. The average difference between the two velocities is about 35%. A contour map of interval velocities computed to an accuracy of 10% should show the shaling up pattern over the area in gross terms. For a 250 m thick interval at an average depth of 2000 m, the corresponding r.m.s. velocity accuracy is about 1·0%. This degree of precision requires very favourable conditions and extremely tight control.

Suppose that, in the above example, a detailed study of facies changes in a much thinner interval was required so that an r.m.s. velocity accuracy of 0·2% was necessary. This means that an r.m.s. velocity in the region of 2500 m/s has to be known to better than 5 m/s. Clearly, to obtain such precision and maintain it over the area of study requires near-ideal conditions that are very rarely, if ever, attained in practice. The estimation of sandstone–shale ratio which is, sometimes, attempted in regions of thick clastic series[60] is subject to similarly stringent considerations.

Velocity Spectra Displays
These displays contain redundant information of potential practical application. For example, in a contoured display (e.g. Fig. 10) the character of the coherency build-up associated with a particular reflector is sometimes

used to complement seismic reflection character as evidence for correlation across faults and prediction of facies changes.[61]

Overpressured Zones
Abnormal fluid pressures give rise to anomalously low interval velocities. The prediction of overpressured zones can sometimes be aided by CDP-derived velocities. The normal procedure is to construct a plot of velocity (r.m.s., average or interval) vs depth (or time). Distinct departure of the curve from the normal trend of increasing velocity with depth, towards low velocity, indicates the top of the overpressured zone.[62] A similar effect is sometimes observable directly on the velocity spectra display. Generally, reliable overpressure prediction requires adequate well control.

Depth Conversion
Ray-Tracing Migration
In carrying out direct depth conversion of reflection times by ray-tracing migration, the section is divided into intervals separated by the mapped horizons. In cases where the velocity in each interval is based on velocity analysis data (in contrast to well data) the velocity cannot be varied as a continuous function of depth (or time) within the interval. Accuracy requirements vary between individual problems but, generally, high accuracies would be necessary because the error is cumulative with successively deeper interfaces.

Conversion with Average Velocities
Depth conversion is, in many cases, carried out simply by multiplying the average velocity by the one-way vertical time to the reflector. The average velocity values may be estimated from r.m.s. velocities or time averaged interval velocities (see the section on average velocity). The method of depth conversion of reflection times on seismic sections is self-evident. For converting time contour maps or uncontoured time values on a map, an average velocity contour map is usually prepared from values derived from MCS velocity measurements along individual profiles.

The accuracy of the average velocity values, whether on a section or on a map, should be sufficient to bring out the mapped structural anomaly well above the level of uncertainty. Considerable care in correcting for various errors would, normally, be necessary. The removal of statics-generated velocity oscillations (see the section on lateral resolution and statics) is usually an important task in the process; the presence of any residual oscillations could produce unreal structural anomalies. The final average velocity map should also be consistent with data from any wells in the area.

ACKNOWLEDGEMENTS

I am grateful to Dr P. N. S. O'Brien for providing BP Research Centre facilities. I am indebted to British Petroleum Exploratie Maatschappij Nederland B.V., Gulf Oil Exploration and Production Co., BP Petroleum Development Ltd, Seiscom-Delta Inc., and Geophysical Prospecting, for kindly permitting the use of their data. I thank the Chairman and Board of Directors of the British Petroleum Co. Ltd for their permission to publish this work.

REFERENCES

1. GREEN, C. H., *Geophysics*, **3**, p. 295, 1938.
2. TELFORD, W. M., GELDART, L. P., SHERIFF, R. E. and KEYS, D. A., *Applied geophysics*, p. 352, Cambridge Univ. Press, Cambridge, 1976.
3. DIX, C. H., *Geophysics*, **20**, p. 68, 1955.
4. MAYNE, W. H., *Geophysics*, **27**, p. 927, 1962.
5. DÜRBAUM, H., *Geophys. Prospecting*, **2**, p. 151, 1954.
6. KREY, TH., *Erdöl Kohle*, **7**, p. 8, 1954.
7. HANSEN, R. F., *Bol. Inform. Petrol.*, **24**, p. 237, 1947.
8. GARDNER, G. H. F., FRENCH, W. S. and MATZUK, T., *Geophysics*, **39**, p. 811, 1974.
9. DOHERTY, S. M. and CLAERBOUT, J. F., *Geophysics*, **41**, p. 850, 1976.
10. TANER, M. T. and KOEHLER, F., *Geophysics*, **34**, p. 859, 1969.
11. SHAH, P. M. and LEVIN, F. K., *Geophysics*, **38**, p. 643, 1973.
12. AL-CHALABI, M., *Geophys. Prospecting*, **21**, p. 783, 1973.
13. SCHNEIDER, W. A. and BACKUS, M. M., *Geophysics*, **33**, p. 105, 1968.
14. AL-CHALABI, M., *Geophys. Prospecting*, **22**, p. 458, 1974.
15. AL-CHALABI, M., *BP Research Centre Report EPR/R*1174, 1973.
16. LUCAS, A. L., *BP Research Centre Report EPR/TN*1077, 1976.
17. CRESSMAN, K. S., *Geophysics*, **33**, p. 399, 1968.
18. LEVIN, F. K., *Geophysics*, **36**, p. 510, 1971.
19. SHAH, P. M., *Geophysics*, **38**, p. 812, 1973.
20. KREY, TH., *Geophys. Prospecting*, **24**, p. 52, 1976.
21. HUBRAL, P., *Geophysics*, **41**, p. 233, 1976.
22. HUBRAL, P., *Geophys. Prospecting*, **24**, p. 478, 1976.
23. SHERIFF, R. E., *Encyclopedic dictionary of exploration geophysics*, p. 228, SEG, Tulsa, Okla, USA, 1973.
24. GAROTTA, R. and MICHON, D., *Geophys. Prospecting*, **15**, p. 584, 1967.
25. NEIDELL, N. S. and TANER, M. T., *Geophysics*, **36**, p. 482, 1971.
26. ROBINSON, J. C., *Geophysics*, **34**, p. 330, 1969.

27. BOIS, P. and LA PORTE, M., *Geophys. Prospecting*, **18**, p. 489, 1970.
28. SHERWOOD, J. W. C. and POE, P. H., *Geophysics*, **37**, p. 769, 1972.
29. AL-CHALABI, M., *BP Research Centre Report EPR/R*1180, 1974.
30. LEVIN, F. K., *Geophysics*, **38**, p. 771, 1973.
31. AL-CHALABI, M., *BP Research Centre Report EPR/TN*1021, 1974.
32. LUCAS, A. L., *BP Research Centre Report EPR/R*1223, 1976.
33. LUCAS, A. L., *BP Research Centre Report EPR/TN*1027, 1974. (See also RENICK, H., with appendix by Lucas, A. L., *Geophys. Prospecting*, **22**, p. 54, 1974.)
34. O'BRIEN, P. N. S. and LUCAS, A. L., *BP Research Centre Report EPR/R*1231, 1977.
35. LINDSEY, J. P., *Geophysics*, **35**, p. 461, 1970.
36. FITCH, A. A., *Seismic reflection interpretation* (eds. Anstey, N. A. and O'Brien, P. N. S.) Gebrüder Bornträger, Berlin, Germany, 1976.
37. BODOKY, T. and SZEIDOVITZ, ZS., *Geoph. Trans. Hungarian Geoph. Inst.*, **20**, p. 47, 1972.
38. SCHNEIDER, W. A., *Geophysics*, **36**, p. 1043, 1971.
39. EVERETT, J. E., *Geophys. Prospecting*, **22**, p. 122, 1974.
40. O'DOHERTY, R. F. and ANSTEY, N. A., *Geophys. Prospecting*, **19**, p. 430, 1971.
41. SPENCER, T. W., EDWARDS, C. M. and SONNAD, J. R., *Geophysics*, **42**, p. 393, 1977.
42. LEVIN, F. K., *Geophysics*, **42**, p. 1043, 1977.
43. DUNKIN, J. W. and LEVIN, F. K., *Geophysics*, **38**, 635, 1973.
44. WEATHERBY, B., BORN, W. T. and HARDING, R. L., *Bull. AAPG*, **18**, p. 106, 1934.
45. BACKUS, G. E., *J. Geophys. Res.*, **67**, p. 4427, 1962.
46. UHRIG, L. F. and VAN MELLE, F. A., *Geophysics*, **20**, p. 774, 1955.
47. THOMAS, J. H., *BP Research Centre Report EPR/TN*1068, 1976.
48. VAN DER STOEP, D. M., *Geophysics*, **31**, p. 900, **31**, 1966.
49. SPENCER, T. W. and BHAMBANI, D. J., *Geophysics*, **40**, p. 426, 1975.
50. LEVIN, F. K. and SHAH, P. M. *Geophysics*, **42**, p. 957, 1977.
51. DINSTEL, W. L., *Geophysics*, **36**, p. 415 (with editorial comment, F. K. Levin), 1971.
52. TOOLEY, R. D., SPENCER, T. W. and SAGOCI, H. F., *Reflection and transmission of plane compressional waves*, AD429, US Dept of Commerce, 1963.
53. AL-CHALABI, M., *BP Research Centre Report EPR/TN*1012, 1973.
54. MILLER, M. K., *Geophysics*, **39**, p. 427, 1974.
55. TANER, M. T., KOEHLER, F. and ALHILALI, K. A., *Geophysics*, **39**, p. 441, 1974.
56. SHAW, J. L., *BP Research Centre Report EPR/R*1193, 1974.
57. TANER, M. T., COOK, E. E. and NEIDELL, N. S., *Geophysics*, **35**, p. 551, 1970.
58. FRENCH, W. S., *Geophysics*, **40**, p. 961, 1975.
59. TUCKER, P. M. and YORSTON, H. S., *Pitfalls in seismic interpretation*, SEG, Tulsa, Okla, USA, 1973.
60. TEGLAND, E. R., *23rd Annual Midwestern Regional Meeting*, SEG and AAPG preprint, Dallas, Texas, USA, 1970.
61. COOK, E. E. and TANER, M. T., *Oil Gas J.*, **67**, p. 159, 1969.
62. PENNEBAKER, E. S., JR., *Symp. Geophysical Society of Houston*, SPE preprint 2165, 1969.

Chapter 2

PATTERNS OF SOURCES AND DETECTORS

S. D. Brasel
Seismic International Research Corporation, Denver, Colorado, USA

SUMMARY

Patterns are the application of array theory and they spatially filter organised interference. The arrangement of source and receiver elements, traces of a seismic record, and common depth point (CDP) trace gathers, form patterns.

The Fourier transform of a pattern produces an array or filter response curve. Where the noise occupies the same frequency band as the signal, but the noise wavelengths differ from those of the signal, an array can be designed such that it will accept the signal and attenuate the noise.

The results to be derived from array theory are not realised in practice as they should be, because of poorly designed experiments and incorrect conclusions drawn from them. The theory works; it is the application that needs improvement.

TYPES OF PATTERN

Patterns vary by numbers of elements, positions, and distances between positions, giving three schemes:

Linear	Element weighted	Distance weighted
○○○○○○○○	(clustered pattern)	○ ○ ○ ○○○○ ○ ○ ○
1 1 1 1 1 1 1 1	1 2 3 4 4 4 3 2 1	1 1 1 1 1 1 1 1 1 1

Each scheme typifies a family of patterns, with certain physical traits and limitations.

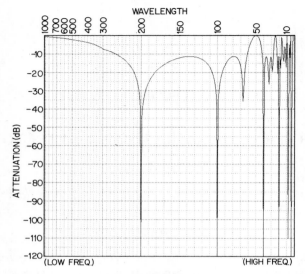

FIG. 1(a). Four elements at 50 m spacing (1 1 1 1).

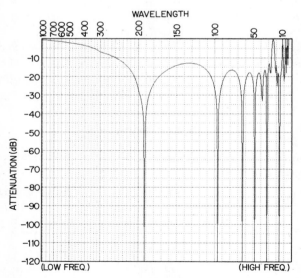

FIG. 1(b). Eight elements at 24 m spacing (1 1 1 1 1 1 1 1).

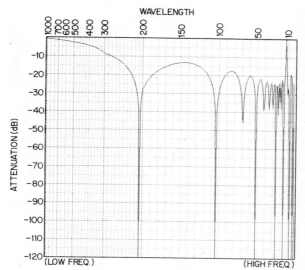

FIG. 1(c). Sixteen elements at 13 m spacing (1 1 1 1 1 1 1 1 1 1 1 1 1 1 1 1).

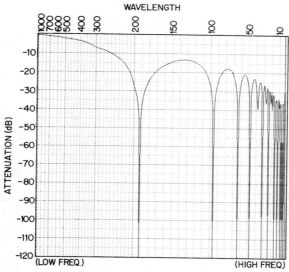

FIG. 1(d). Sixty-four elements at 3 m spacing (1 1 1 ... 64 linear ... 1 1).

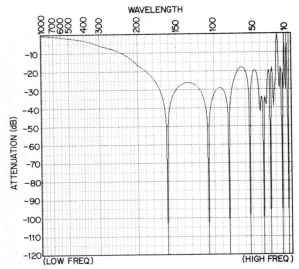

FIG. 2(a). Eighteen elements weighted, spacing 18 m (1 1 1 2 2 2 2 2 2 1 1 1).

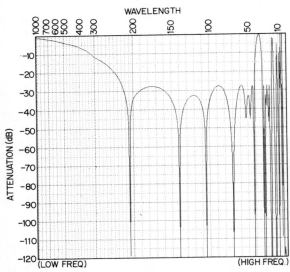

FIG. 2(b). Twenty-four elements weighted, spacing 35 m (1 2 3 4 4 4 3 2 1).

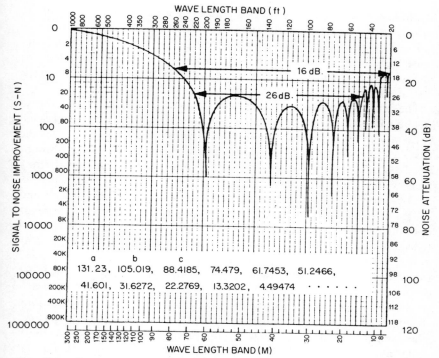

FIG. 2(c). Twenty-two elements distance weighted, spacing 131·23, 105·019, 88·4185, 74·479, 61·7453, 51·2466, 41·601, 31·6272, 22·2769, 13·3202, 4·4974 m.

Linear patterns are the bandwidth application of array theory. The bandwidth (the ratio of the longest to shortest wavelength in the attenuation zone) is approximately equal to the number of elements in the linear array. Increasing the number of elements beyond about eight extends the bandwidth but does not appreciably change the power of attenuation.

Figures 1(a), (b), (c) and (d) are response curves for linear arrays. Beyond eight elements the bandwidth increases but not the power of attenuation. Figures 2(a), (b) and (c) are examples of element and distance weighting.

PATTERN SELECTION

Pattern selection is a very involved process, incorporating almost every facet of seismology. The objective is simply to attenuate noise and hence improve the signal–noise ratio of a recording. Selecting signal enhancement procedures depends upon the expected physical environment creating the signal and noise frequency spectra and wavelengths measured along the surface in the direction of the spread. Patterns discriminate by spatial filtering, therefore where signal and noise frequencies overlap, the improvement in signal–noise ratio depends on their difference in apparent velocity, and hence in wavelength, along the line of traverse.

The first consideration is to select the frequency band. This is not always easy as many physical parameters of source, receiver and geological environment must be understood. Ideally broad-band recording gives the best seismic resolution and should be initially planned, even though broad-band recordings may require expensive field operational techniques, or may even be impossible to achieve.

The second consideration is to select the bandwidth of the receiver and source arrays to be used in the field technique. This parameter depends upon the velocity range of the seismic noise family, both direct and reflected. The velocity range and the frequency range can be measured by noise profiles or estimated for a geological environment; the peak frequency can be estimated by timing from trough to trough on the noise events, or by spectrum analysis of short windows dominated by noise.

The third consideration is to calculate the apparent wavelengths along the spread of the noise to be attenuated and the signal wavelets to be passed. These depend on the range of apparent velocities and frequencies of both signal and noise. The reflector wavelengths depend on dip and normal moveout. A preliminary calculation can be based on the moveout of a refraction in the bed below the shallowest refractor. Usually, the objective of a seismic survey is quite specific, and is represented by deeper reflectors, with a much smaller moveout. (*Note:* the direct wave would travel at a surface wave velocity with a large moveout; in practice the asymptote to the moveout hyperbola is not observed—it passes into a straight line refractor velocity which is tangential to the hyperbola.)

The fourth and most varied consideration is to select the physical techniques, and it depends upon the many parameters associated with generating, receiving, and recording seismic data. These include geologically controlled source and receiver effects, the seismic field equipment and, probably the most critical, the operational efficiency and economics. Advisably several techniques of varying power and cost should be designed, tested and later selected on the basis of resolving power versus cost. Low power and hence low cost procedures may perform satisfactorily for reconnaissance surveys but detailed well site selections and local 'severe' noise areas will probably require stronger procedures.

The calculation of an array response assumes uniform conditions at the surface of the earth, precise placement of sources and detectors, and uniform output from sources and detectors. In practice, departures from these and other ideal conditions, especially local changes in elevation and weathering within the pattern, introduce error into the waveforms at individual geophones which should, ideally, be identical. The impact of these errors on the performance of a pattern must be considered in making a final selection. The patterns critically sensitive to the theoretical null points will not perform well in the field. The null points do not appear in practical field usage. Highly sensitive arrays, those of a few positions, or strongly weighted (either distance or element) quickly lose their theoretical advantages. Pattern effectiveness improves with the multiplicity of positions relative to weighting functions.

The ambient noise attenuation power of patterns refers to their effectiveness against random noise, although few ambient noise problems are random. If a noise source is close to a geophone, and of low power, its output will be recognised only in the output of that geophone. A more distant noise source will be recorded on several geophones, if it is of sufficient power, and so it will appear as aligned or organised noise. There is a close relationship between the statistical properties of patterns and ambient noise attenuation power. The random variables of distance and weighting accuracy are similar to random noise distribution of ambient noise and source and receiver coupling variations. The basic consideration for effectiveness against random variables is:

$$\left.\begin{array}{l}\text{signal–noise (S–N) improvement}\\ \text{(dB) if noise is random}\\ \text{(for weighted samples)}\end{array}\right\} = 20\log_{10}\frac{\sum\limits_{i=1}^{n} w_i}{\left(\sum\limits_{i=1}^{n} w_i^2\right)^{1/2}}$$

where w_i is the weight of the ith sample and n is the number of samples, or signal–noise (S–N) improvements, (dB) (for equally weighted samples) signal–noise (S–N) improvement (dB) = $10\log_{10}$ (multiplicity).

Example

Dispersed weighted array

1 2 3 4 4 4 3 2 1

S–N improvement = $10\log 24$ = 13·8 dB

Weighted array 1 2 3 4 4 4 3 2 1

$$\text{S–N improvement} = 20 \log \frac{24}{(76^2)^{1/2}} = 8\cdot 8\,\text{dB}$$

The dispersed array is clearly 5 dB better for random noise. In the foregoing treatment the statistical assumption made is that the signals at each source or detector are in phase, and that the noise is random, or is unique at each array position.

Array theory applications must be given consideration in all phases of field acquisition and processing of seismic data, from the simplest single shot to the most complex computer modelling program. Traditionally, patterns have been used only for land ground roll, but recent applications to water bottom multiples demonstrate their wide application to organised interference.

Water bottom multiples, for example, are peaked in the frequency spectrum, due to multiple occurrences and the effect of the bubble pulse. They are also peaked in the wavelength spectrum, if distant traces are considered, where the multiples' moveout curves have asymptotically approached water velocity. The bubble pulse is frequently set to occur with 35 ms delay, thereby peaking the unwanted signals of about 150 ft wavelength. Fabricating the source, receiver, and CDP arrays to notch filter the peaked water bottom events in the wavelength domain attenuates the sea bottom multiples and so improves the geological resolution. Similarly, random code source systems, operating on a moving boat, can be tuned to suppress lower velocity systems such as water bottom multiples.

Shot-hole dynamite techniques require special analyses. Buried charges spatially filter primary ground roll by the following theory: wavelengths equal to shot depth are attenuated at 19 dB, twice the hole depth wavelengths at 6 dB, but wavelengths half the hole depth are attenuated at 35 dB. Therefore when selecting receiver arrays for dynamite crews the buried charge effect should be acknowledged. The buried charge ground roll attenuation applies to only the primary or initial shot effect, subsequent noise waves such as 'flexural' energy from near surface rigid layers will not be attenuated, and must be handled by source and receiver arrays.

An Example of Array Technique

Assume the geological near surface to be surface clay with compressional wave velocity, $V = 4250\,\text{ft/s}$, and overlying shale with $V = 7750\,\text{ft/s}$. Surface waves are generated in each of these surface formations at velocities which are about 40 % of these compressional wave velocities, so that the ground roll velocities can be written $V_{\text{high}} = 3100$ and $V_{\text{low}} = 1700$.

Estimated ground roll velocity range, V_R

$$= \frac{V_{high}}{V_{low}} = \frac{40\% \times 7750}{40\% \times 4250} = \frac{3100}{1700} = 1 \cdot 83$$

Bandwidth of field patterns BW_{FIELD} (direct noise), with ideal broad-band recording, 8–62 Hz

$$= BW_{SIGNAL} \times V_R = \frac{62}{8} \times 1 \cdot 83 = 14 \cdot 18$$

with practical field or processing band recording, 16–48 Hz

$$= \frac{48}{16} \times 1 \cdot 83 = 5 \cdot 5$$

The difference in the design of arrays for broad-band and lesser bandwidths is the number of elements required to gain a level of signal enhancement. Wavelength bands (direct noises) for longest and shortest wavelength are:

(1)   Longest wavelength, $\lambda_L = \dfrac{\text{Highest noise velocity}}{\text{Lowest frequency}}$

   Broad-band $\lambda_L = \dfrac{3100}{8} = 387 \cdot 5$ ft

   Practical band $\lambda_L = \dfrac{3100}{16} = 193 \cdot 8$ ft

(2)   Shortest wavelength, $\lambda_S = \dfrac{\text{Lowest noise velocity}}{\text{Highest signal frequency}}$

   Broad-band $\lambda_S = \dfrac{1700}{62} = 27$ ft (assuming no air waves)

   Practical band $\lambda_S = \dfrac{1700}{48} = 35$ ft

Hence, for direct noise waves broad-band = 27–387·5 ft and practical band = 35–193·8 ft.

For direct waves, a range of choices of pattern and hole depth is possible to achieve a required level of attenuation over a given range. For example, one can use shallow holes to attenuate the shorter wavelengths in the reject band, and long patterns to attenuate the longer wavelengths in that band. Alternatively, one can use deep holes to deal with the longer wavelengths, and short patterns to deal with the shorter wavelengths.

Reflected noise often produces poor data. It presents a special problem because it arrives later than the direct noise arrival, and so causes interference at a time on the recording which would otherwise be free from noise. The reflected noise can thus obscure the arrivals from deep horizons. Surface waves can be strongly reflected at surface lithological features or topographic features. Moreover, the amplitude of these surface waves decays slowly with time; not only because they travel with low velocity (and amplitude loss is a function of distance travelled) but also because surface waves are subject to circular spreading, rather than the spherical spreading of compressional waves from deep reflectors. The result is that the interference from this reflected noise may be of especially high relative amplitude.

The velocity of a reflected noise wave is identical to that of the direct wave at the same time, when velocities are measured in the direction of propagation. However, the reflected noise wave often arrives obliquely to the line of the spread, and so it will have a higher apparent velocity. In the limiting case, where the surface reflecting features are nearly parallel to the spread, the noise wave will be broadside to the spread, and the apparent velocity will approach infinity at some points on the spread. Since the surface noise wave has a low velocity, it will arrive at the spread with a strong moveout hyperbolic curvature.

Somewhat empirically, these oblique reflected waves can be dealt with partially by using an increased apparent velocity which is 1·5 times that estimated from the compressional wave velocity. This leads to an increase of 1·5 times in the value V_R.

Reflected Noise Areas
$$BW_{FIELD} = BW_{SIGNAL} \times V_R \times 1.5$$
For the example;
with broad-band recording (reflected noise area)

$$BW_{FIELD} = \frac{62}{8} \times 1.83 \times 1.5 = 21.27$$

and with practical band recording

$$BW_{FIELD} = \frac{48}{16} \times 1.83 \times 1.5 = 8.25$$

The field and processing band (reflected noise areas) for 8–62 Hz filter is 27–562 ft and for 16–48 Hz filter is 35–281 ft.

Broad-band recording is very desirable but leads to expensive field operations. The example requires twenty elements, source and receiver, just to reach bandwidth, and this only allows linear 12 dB patterns; a weak conventional signal enhancement technique.

The data processing bandwidth, determined by computer analysis, is frequently severely restricted, but agrees with the theoretical response curves of the actual field operation. Knowing the frequency band of field patterns automatically decides processing parameters.

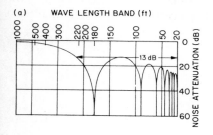
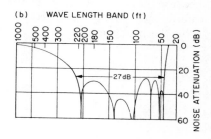
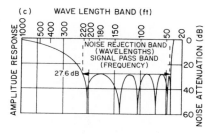

FIG. 3. Optimisation of seismic field techniques; (a) linear systems, (b) poorly designed techniques and (c) optimised field systems.

The efficiency of array attenuation is easily observed by inspection of the response curve. The optimum array has equal nodes or peaks within the attenuation band. If a pattern has radically varying peak values then the attenuating threshold will be improved by rearrangement of the elements or the same attenuation can be achieved with fewer elements. Figure 3 compares two poorly designed thirty-six geophone patterns to a twenty-four unit optimum array.

Adapting field crew equipment for varied field problems needs considerable computer analysis. For any set of equipment all possible physical patterns should be calculated and grouped by bandwidth and attenuating power, ready for instant field testing and application. A Fortran program, not requiring an array processor, is included in the appendix for general adaptation. Also included is a set of practical field patterns optimised for a 5·5 bandwidth easily utilised for geophones and surface source systems.

Comparisons of field techniques are included in the appendix to demonstrate that array theory works. Where the reflection from a geological horizon cannot be observed on the seismic section it is usually because noise is arriving at the same time as the reflection. When interference is analysed correctly a simple array theory approach will often solve the problem.

APPENDIX 1: PRACTICAL ARRAYS OPTIMISED FOR BANDWIDTH ≈ 5·5

	Array	Number of elements	Bandwidth	Attenuation
A	1 1 1 1 1 1 1 1 1 1 1 1 1 1 1 1 ───────────────── 1 1 2 2 2 2 2 2 1 1	18	10·0	18 dB

B	1 2 2 2 2 1
	    1 2 2 2 2 1
	─────────────────
	  1 2 3 4 4 3 2 1
		20	4·6	25 dB

C	1 2 2 2 2 2 1
	    1 2 2 2 2 2 1
	─────────────────
	  1 2 3 4 4 4 3 2 1
		24	5·5	27·6 dB

D	1 2 3 4 4 3 2 1
	  1 2 3 4 4 3 2 1
	    1 2 3 4 4 3 2 1
	─────────────────
	1 2 4 6 8 9 9 8 6 4 2 1
		60	5·6	35 dB

E	1 2 3 4 5 5 4 3 2 1
	  1 2 3 4 5 5 4 3 2 1
	    1 2 3 4 5 5 4 3 2 1
	      1 2 3 4 5 5 4 3 2 1
	─────────────────
	1 3 6 10 14 17 18 17 14 10 6 3 1
		120	5·6	44 dB

F	1 2 3 4 5 5 5 5 4 3 2 1
	  1 2 3 4 5 5 5 5 4 3 2 1
	    1 2 3 4 5 5 5 5 4 3 2 1
	      1 2 3 4 5 5 5 5 4 3 2 1
	        1 2 3 4 5 5 5 5 4 3 2 1
	          1 2 3 4 5 5 5 5 4 3 2 1
	or
	1 2 4 6 8 9 9 8 6 4 2 1
	  1 2 4 6 8 9 9 8 6 4 2 1
	    1 2 4 6 8 9 9 8 6 4 2 1
	      1 2 4 6 8 9 9 8 6 4 2 1
	─────────────────
	1 3 6 10 15 20 24 27 28 27 24 20 15 10 6 3 1
		240	5·7	50 dB

APPENDIX 2: DINOSEIS ARRAY TESTS USING MICROSPREAD, CANADIAN ARCTIC

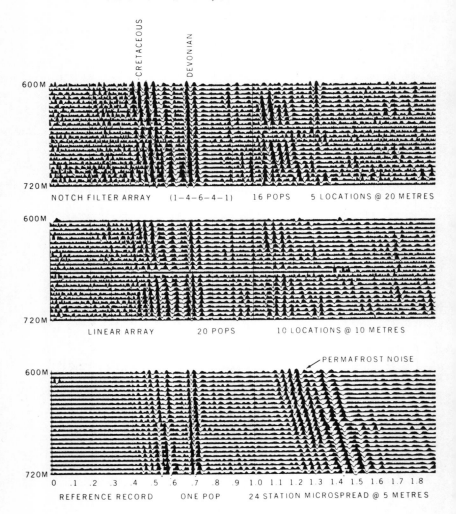

APPENDIX 3: SUB-THRUST REEF PLAY, CANADA 1969

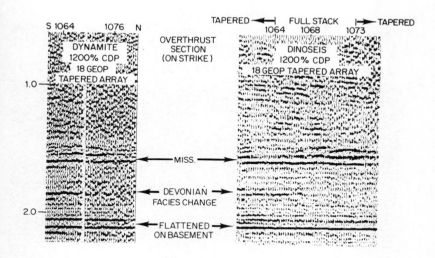

APPENDIX 4: FIELD TECHNIQUE COMPARISON, MEXICO

Top; original technique, 180 pops (1, 1, 1, 2, 2, 2, 2, 2, 2, 1, 1, 1) 24 geophones (1, 1, 1, ... linear). Bottom; improved technique, 60 pops (1, 3, 5, 7, 9, 10, 9, 7, 5, 3, 1) 24 geophones (1, 1, 1, 1, 2, 2, 2, 2, 2, 2, 2, 2, 1, 1, 1, 1).

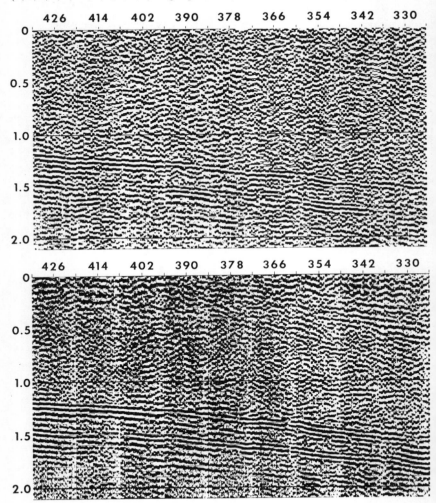

APPENDIX 5: CIRCULAR NOISE PROFILE, POLAND 1973

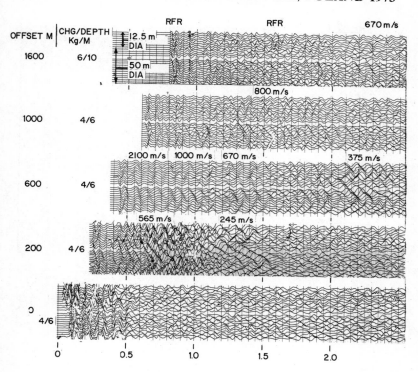

APPENDIX 6: NOISE STUDY, POLAND 1973

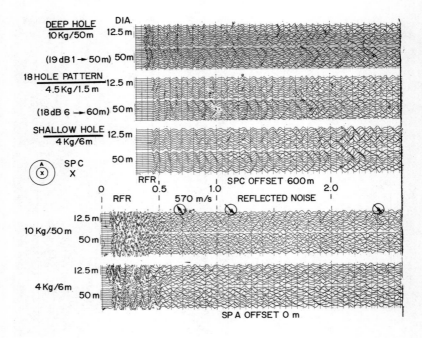

APPENDIX 7: ORIGINAL DYNAMITE TECHNIQUE, POLAND

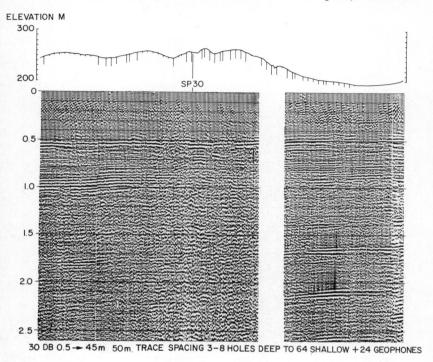

APPENDIX 8: NEW DINOSEIS TECHNIQUE USING DYNAMITE, POLAND

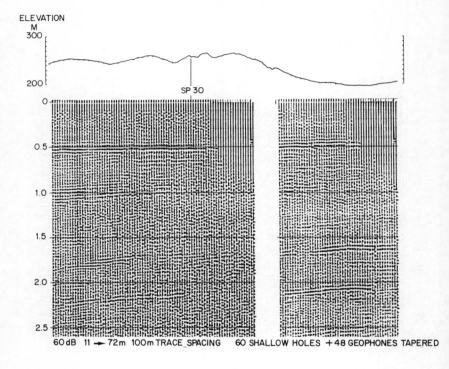

APPENDIX 9: FORTRAN PROGRAM; PATTERN RESPONSE CURVES

```
      DIMENSION A( 600),AS( 600)
      CC = 1.OE-30
  101 FORMAT(1H,8X,24F3.0)
  301 FORMAT(35F2.0)
   20 DO 50 I = 1,600
      A(I) = 0.
      AS(I) = 0.0
      B(I) = 0
   50 C(I) = 0
      READ(2,102)N1,L1
  102 FORMAT(215)
      IF(N1)210,999,210
  210 READ(2,301)(A(I),I = 1,N1)
      WRITE(5,104)
  104 FORMAT(1H1,8X,'FIRST ARRAY VALUES')
      WRITE(5,101)(A(I),I = 1,N1)
      ISTAN = (N1-1)*L1
      WRITE(5,103)N1,L1,ISTAN
  103 FORMAT(1H,8X,'NUMBER OF POINTS',I5,10X,
     'SPACING',I5,10X,'LENGTH',I7)
      J1 = (L1*(N1-1)) + 1
      J = 1
      DO 2 I = 1,J1,L1
      C(I) = A(J)
    2 J = J + 1
      READ(2,102)N2,L2
      IF(N2)100,100,3
    3 READ(2,301)(B(I),I = 1,N2)
      WRITE(5,106)
  106 FORMAT(1HO,8X,'SECOND ARRAY VALUES')
      WRITE(5,101)(B(I),I = ,N2)
      ISTAN = (N2-1)*L2
   34 WRITE(5,103)N2,L2,ISTAN
      J2 = (L2*(N2-1)) + 1
      IF(J1 + J2-610 )48,48,47
   47 WRITE(5,107)
  107 FORMAT('TOTAL SPREAD TOO BIG FOR THIS PROGRAM')
      GO TO 20
```

```
   48 CONTINUE
      DO 4 I = 1,J2
    4 A(I) = 0
      J = 1
      DO 5 I = 1,J2,L2
      A(1) = B(J)
    5 J = J + 1
      J3 = J1 + J2-1
      DO 6 I = 1,J3
    6 B(I) = 0
      DO 9 I = 1,J1
      DO 9 J = 1,J2
      K = 1 + J-1
    9 B(K) = C(I)*A(J) + B(K)
      DO 8 I = 1,J3
    8 A(I) = B(I)
      N = J3
      GO TO 10
  100 N = N1
      DO 49 I = 1,J1
   49 A(I) = C(I)
   10 PI = 3.14159265
      P12 = PI*2.
      J = N/2
      D = 1400.
      DO 21 I = 2,5
      D = D-200.
   21 B(I) = 1./D
      DO 22 I = 6,11
      D = D-50.
   22 B(I) = 1./D
      DO 23 I = 12,21
      D = D-10.
   23 B(I) = 1./D
      DO 24 I = 22,41
      D = D-5.
   24 B(I) = 1./D
      DO 25 I = 42,90
      D = D-2.
```

```
25  B(I) = 1./D
    B(1) = 0
    DO 12 I = 1,90
    XH = B(I)
    XB = P12*XH
    XA = COS(XB)
    XC = SIN(XB)
    EX = A(J + 1)
    EY = 0.
    XE = XA
    XF = XC
    DO 13 K = 1,J
    J1 = J + 1
    J2 = J1 + K
    J3 = J1-K
    EX = EX + (A(J2) + A(J3)*XE
    EY = EY + (A(J2)-A(J3)*XF
    XG = XE*XA-XF*XC
    XF = XF*XA + XE*XC
13  XE = XG
    C(I) = SQRT(EX*EX + EY*EY)
12  CONTINUE
    AMAX = C(I)
    DO 14 I = 1,90
    IF(AMAX-C(I))15,14,14
15  AMAX = C(I)
14  CONTINUE
    ANOR = 1./AMAX
    WRITE(5,108)ANOR,AMAX
108 FORMAT (1H0,8X,'NORMALISATION FACTOR =',
    F8.5,10X,'MAXIMUM AMPLITUDE =',F5.0)
    WRITE(5,200)
200 FORMAT (1H0,8X,'WAVELENGTH AMPLITUDE AMP(DB)',
    8X,'WAVELENGTH AMPLITUDE AMP(DB)')
    DO 16 I = 2,90
    C(I) = C(I)*ANOR
    IF(C(I)-CC)500, 501,501
500 C(I) = CC
501 AS(I) = 20.*(ALOG(C(I))*0.4342945)
    B(I) = 1./B(I)
```

```
 16 CONTINUE
    WRITE (5,110)(B(I),C(I),AS(I),B(I+45),C(I+45),AS(I+45),I=2,45
    WRITE(5,111) B(46),C(46),AS(46)
    WRITE(5,125)
125 FORMAT  (1H,'COURTESY  STAN  BRASEL,  SEISMIC
    INTERNATIONAL RESEARCH)
    GO TO 20
999 CALL EXIT
110 FORMAT (8X,F10.0,2F10.4,4X,F10.0,2F10.4)
111 FORMAT (8X,F10.0,2F10.4)
    END
```

Chapter 3

WELL GEOPHONE SURVEYS AND THE CALIBRATION OF ACOUSTIC VELOCITY LOGS

P. KENNETT

Seismograph Service (England) Limited, Keston, Kent, UK

SUMMARY

Very few exploration wells are now drilled without some kind of velocity measuring technique being included in the logging programme, and, increasingly, development and appraisal wells are also being used for velocity determinations. Many parameters of interest to both exploration and exploitation staff can be inferred or calculated from logged or measured velocity data, such as,

(1) *velocity distribution with depth for seismic data control,*
(2) *synthetic seismograms for reflection analysis and correlation,*
(3) *lithological change and structural information,*
(4) *change in fluid content and porosity determination,*
(5) *recognition of over-pressure zones.*

All of these uses require good control in the data acquisition phase, and an understanding of the basic principles of acoustic wave propagation. Limitations inherent in the methods of acquisition, analysis and interpretation should be carefully weighed, as their influence may vary according to the purpose for which the data are utilised.

INTRODUCTION

Until the mid 1950s, when the first continuous acoustic velocity logs appeared commercially, the most direct way to measure seismic velocities in

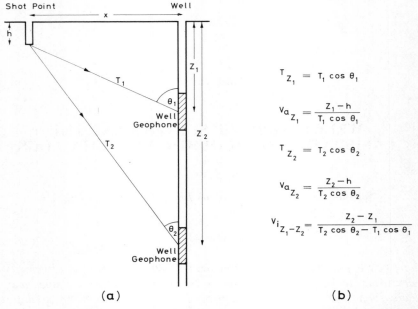

FIG. 1. (a) Schematic of conventional well geophone survey, and (b) derivation of interval and average velocities.

the deep subsurface was to measure the travel-time of a pulse generated by an explosive charge at the surface and detected by a pressure sensitive transducer at various fixed depths in a well. This is shown schematically in Fig. 1(a). After suitable corrections were applied to the observed times for the depth of the shot and its horizontal offset from the well, a series of average and interval velocities for each detector position was calculated (Fig. 1(b)). In the general case, the calculations assume that the Fermat ray path is a straight line between shot and detector; that is to say, Snell's law is not obeyed at the many discontinuities in velocity and density between surface and detector.

This technique remained in general use, both on shore and off shore, until about 1970 using dynamite as the explosive source of seismic energy, using pressure sensitive seismometers down-hole, which required complicated pressure compensation devices to offset the considerable and variable hydrostatic borehole pressures, and using paper or film cameras to record the shot instant and well geophone signals. The method was generally satisfactory for land wells where rig-time was relatively inexpensive,

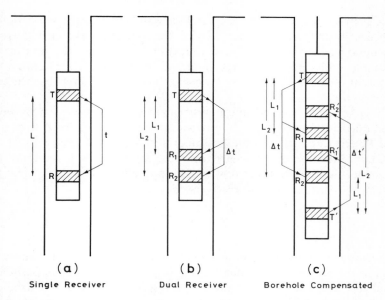

FIG. 2. Schematic of geometry of some common acoustic logging sondes.

explosives could be used safely and conveniently, signal to noise problems could be largely overcome by choice of charge size, and where first arrival times of the seismic pulse at the detector was the major geophysical consideration.

The period between the mid 1950s and 1970 saw the introduction and development of the continuous log of acoustic travel-time variously known commercially as the Continuous Velocity Log (CVL), Acoustilog and Sonic Log. The principle of this log is to generate acoustic energy, with a wavelength greater than the pore and grain size of most sedimentary rocks, and observe, in the same borehole instrument, its travel-time over a fixed short distance of borehole wall. These transit times are then summed continuously with respect to depth so that a total travel-time measurement may be observed. There have been three distinct stages of development of this device until the present time, which are illustrated schematically in Fig. 2.

In Fig. 2(a) the original single transmitter (the single receiver logger) is shown, in which only one transit time measurement was made for the passage of acoustic energy between T and R over the vertical distance L. Note that this time included two segments of travel path in the borehole fluid between the sonde and the borehole wall, and that the separation L between T and R was in fact greater than the vertical travel path in the rock surrounding the borehole due to refraction of the acoustic wave at the borehole walls. Because of the influence of the mud travel paths, the apparent velocity given by $V = L/t$ usually gave values lower than the true formation velocity.

In order to minimise the effect of the mud travel paths, a second receiver was added as shown in Fig. 2(b); two transit times were observed between T and R_1 and T and R_2, over the vertical distances L_1 and L_2 respectively. By assuming that the mud travel paths were equal for both receivers, the difference in observed times δt was used to calculate the velocity over the distance separating the two receivers, viz.

$$V = \frac{L_2 - L_1}{\delta t}$$

Because the assumption of equal mud travel times is invalid in many wells (e.g. in the presence of irregular caving of the borehole wall opposite the sonde, or where the sonde is tilted with respect to the borehole axis) the configuration of two transmitters and four receivers was introduced, as shown in Fig. 2(c). Fast switching of the two transmitter–receiver arrays allows two interval transit times to be observed, δt and $\delta t'$, over the same receiver spacing. These times are then averaged to determine the velocity, viz.

$$V = \frac{\dfrac{L_2 - L_1}{\delta t} + \dfrac{L_2 - L_1}{\delta t'}}{2}$$

The interval transit times, and hence velocities, given by the borehole compensated system are more nearly correct than either of the other two, but are not sufficiently accurate for most seismic processing and interpretation applications. It remains necessary therefore to calibrate these measurements against some other reference, usually the well geophone survey.

The effect of the distances between transmitters and receivers, the so-called span and spacing, needs to be considered in the context of the use to which the velocity data are to be applied. When these distances are small only the altered zone around the borehole will be investigated; to examine the unaltered formation away from the borehole, larger distances are required.

DEVELOPMENTS IN EQUIPMENT AND TECHNIQUE

Most of the developments since 1970 have been by far in the acquisition and processing of the well geophone data. Spurred on largely by the rapid increase in the number of off-shore wells drilled, and the massive increase in rig-time costs, it became necessary to develop an all-weather day or night field technique. This almost automatically ruled out the use of explosives as a seismic energy source and has resulted in the widespread use of air-guns as an alternative. Because of the limited and comparatively low level energy output of air-guns, means of improving signal to noise conditions downhole and some means of summing signals either in-field or elsewhere became mandatory. These criteria have led to the use of wall-clamped well geophones and magnetic tape recording in addition to the conventional paper or film camera.

A number of different air-guns are available commercially and all are based on the principle of containing a fixed volume of compressed air or gas in a chamber such that it can be released at high expansion rates on command. In some, the air is allowed to expand rapidly into the surrounding sea water (or other medium if on shore) to form a pulsating bubble, and in others the air is used to drive a piston which creates a void into which the surrounding medium collapses to generate seismic energy (see Figs. 3 (a) and (b)).

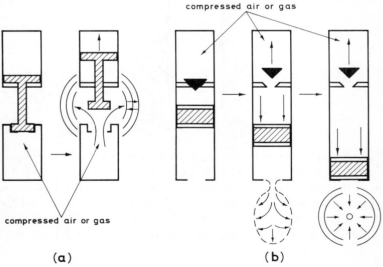

FIG. 3. Schematic of some common air-guns.

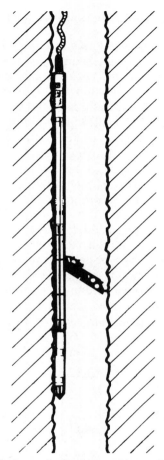

FIG. 4. Wall-clamped velocity sensitive well geophone.

The gun output waveform signature can be controlled to a large extent by the selection of pressure, chamber size, depth and by using guns either singly or in simple arrays. The signature is usually monitored by a pressure sensitive hydrophone located within 2 or 3 m of the gun system, and this is used for zero time reference purposes and other signal control applications. Other sources, such as Vibroseis®, weight drop and other impact sources,

® Trade mark of Continental Oil Co.

are occasionally used on shore, but in general they have no technological advantages and may add to the costs of data acquisition.

The well geophone is the heart of the entire well velocity survey method. Modern geophones are velocity sensitive and designed to clamp to the borehole wall by means of a motor driven arm mechanism, as shown in Fig. 4. This allows the logging cable, by which the geophone is suspended, to be slackened off so that it no longer supports the weight of the geophone and thereby instantly removes one major transmission path for extraneous noise. The degree of clamping to the borehole wall is an important factor if the geophone response is to be without distortion and proportional to the small amounts of particle displacement of the rock surrounding the borehole. There must be no differential movement, i.e. slippage, between geophone and wall during passage of the seismic event; this means in practice that a lateral force equivalent to about two or three times the weight of the geophone must be supplied by the arm mechanism. Techniques exist for qualitatively observing the degree of coupling in the field, and which allow subsequent spectral analysis of the recorded data. Usually, this requires the application of a consistent small impulsive force to the body of the well geophone, when clamped and unclamped, at each geophone location. The geophone output varies according to the coupling as shown in Fig. 5. Electronically, the geophone system should have a wide flat frequency response, and amplification in the geophone to raise the signal output well above unwanted noise.

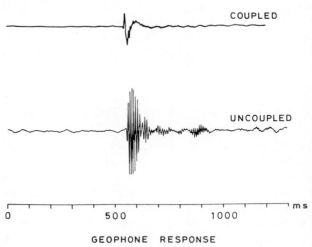

FIG. 5. Recorded signals from coupled and uncoupled wall-clamp geophone.

Magnetic tape recording has become an important addition to the basic field equipment required for well velocity surveys. Its primary purpose is to allow multiple shots to be recorded at each geophone check level so that they can be composited in a computer if required, or for signal enhancement purposes if data is of a low grade. More recently, digital tape recording with in-field summing capabilities, has been introduced so that other signal enhancement and analysis techniques can be applied to the data in addition to the basic time–depth observations.

All of these developments together have improved the acquisition of well geophone data and greatly reduced the amount of time required to perform the surveys. Air-guns can be safely used from the drilling rig platform off shore, so that boats are not required for conventional surveys, and have eliminated the need for shot-holes and explosives on shore. They have allowed the offset distance between source and well to be minimised so that near vertical travel paths to the well geophone can be observed; this has minimised the errors associated with slant travel-times, which were assumed to have followed straight-line paths between shot and detector, and their correction to the vertical.

There have been no major developments in acoustic log field equipment or technique in recent years. Digital tape recording has been introduced for both data transmittal purposes and some computer analysis. Acoustic logs are now often recorded and displayed as one of a multiple string of logs, whereas traditionally they were acquired singly, in order to reduce the number of logging runs required, and hence to minimise rig-time. Some development is under way to produce logging tools with spacings between transmitters and receivers of not less than 3 m, much greater than hitherto, in order to ensure a much larger radius of investigation into the unaltered formation.

THE CALIBRATION OF VELOCITY LOGS

The basic principle involved in the calibration of acoustic logs is to compare the set of times obtained from the well geophone survey with the set of times, referenced to the same depths, obtained from the velocity log; this second set of times is the integration with depth of the interval transit times measured between transmitter and receivers. There are a number of reasons why these two sets of times do not always agree and, in the general case, it is assumed that the differences between them are due to errors in the transit time measurements of the velocity log. The influence of such factors as

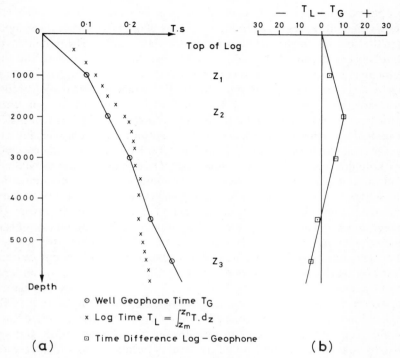

Fig. 6. Comparison of well geophone and acoustic log travel-time data.

lithology and geological structure, which may result in different travel paths for the two types of survey, must however be carefully weighed before adopting this principle rigorously. The most common causes of error in recording the interval transit times are as follows:

(1) Borehole geometrical effects such as caving and tilting of sonde.
(2) Invasion of the borehole wall by drilling fluids, which may be differential with distance away from the wall.
(3) Small radius of investigation of the acoustic signals when span and spacing are 1 m or less; the unaltered rock surrounding the well may not be reached by the signals.
(4) Cycle skipping, caused by attenuation of one or more of the received signals, resulting in inaccurate 'electronic picking' of the signals.

In Fig. 6(a) the time versus depth functions are plotted for the observed geophone times and the integrated log interval transit times. Figure 6(b) shows the differences between these two sets of data as a function of depth, and the best fit straight lines joining the difference values. The plot indicates that between depths Z_1 and Z_2 the log transit times are too long and that velocities determined from them need to be adjusted to higher values to agree with the well geophone data. Below Z_2, velocities from the log need to be decreased. The question for the interpreter is how best to apply these adjustments and at what point to change the direction of adjustment. Z_2 is usually determined by inspection of the log and may or may not fall at a geophone check level. More often than not it will be chosen to coincide with a lithological break, a change in hole diameter, or a change in the incidence and magnitude of caving. In any event, the choice will govern to some extent the slope of the best straight lines through all the difference points, which in turn determines the rate at which time is added to, or subtracted from, the integrated log. The slopes of these straight lines expressed in terms of time per unit distance, usually μs/ft or μs/m, can be applied directly to the log so that every time–depth co-ordinate pair on the log is shifted by this amount. However this may not always be desirable and may not produce a geophysically plausible result. For example, if a layer of salt was present anywhere between depths Z_1 and Z_2 in Fig. 6, linear application of the correction indicated by the slope of the line across that interval would result in the recorded salt velocity being adjusted to a higher value. But salt usually logs 'true', so that no adjustment is necessary, in which case the error indicated by the calibration curve should only be distributed across the non-salt part of the logged interval. In many interbedded geological sequences, such as sand–shale series or evaporite deposits, where velocities may alternate rapidly between high and low values over relatively small intervals of depth, it often happens that the lower velocity components in the series contribute most if not all of the difference between log and geophone times. The distribution of error then needs to be apportioned according to the sedimentation. These considerations thus lead to two different types of correction

$$\varepsilon = \frac{(T_{L_2} - T_{G_2}) - (T_{L_1} - T_{G_1})}{Z_2 - Z_1} \times 1000\,\mu\text{s/ft} \quad \text{or} \quad \mu\text{s/m} \tag{1}$$

This gives a correction to be subtracted from the observed value

$$f = \frac{T_{G_2} - T_{G_1}}{T_{L_2} - T_{L_1}} \times 100\,\% \tag{2}$$

This gives a correction factor by which the observed value is multiplied.

TABLE 1
COMPARISON OF EFFECTS OF LINEAR AND DIFFERENTIAL SHIFTS FOR HIGH AND LOW VELOCITIES

V_L ft/s	δT_L μs/ft	δT_L after correction		Velocities after correction		Change in velocities	
		Linear	Differential	Linear	Differential	Linear	Differential
		μs/ft		ft/s		ft/s	
6 000	167	162	160·5	6 173	6 286	173	286
20 000	50	45	48	22 200	20 800	2 200	800

Equation (1) results in a fixed number of μs/ft or μs/m which can be either added to or subtracted from the transit time curve logged over the depth interval specified in the denominator, and hence represents a linear bulk shift to the curve for that interval, and is independent of velocity. Equation (2) results in a differential shift to the transit time curve since the amount of shift is dependent on the velocities logged. Both equations modify the higher velocities more than the lower, but the differential correction minimises the spread between them. Consider two velocities, 6000 ft/s (1835 m/s) and 20 000 ft/s (6100 m/s) near the lower and upper limits of most common sediments as in Table 1, and that eqns. (1) and (2) give a linear shift of -5 μs/ft and a differential shift of 96·1 % respectively.

Thus in order to minimise the over correction of higher velocities when it is clear that the lower velocity materials surrounding the borehole have contributed most of the error, a differential correction would be used. It is usual to apply such a correction only when $f < 100\%$ and over logged sections where there are large variations of velocity. A refinement of the technique allows a velocity base line to be established above which no shift to the log will be permitted, so that all the error must be distributed to those velocities below the line. This has important consequences in the preparation of synthetic seismograms from the calibrated log, since significant distortion of the reflectivity log may occur if the log errors are not carefully handled. For synthetic seismogram purposes, careful editing of the field velocity log before calibration is an important step. Some of the factors which distort the velocity log, e.g. cycle skipping and caving effects, can be effectively removed before the process of time difference analysis is started. Editing of the log in this way requires skill, and where possible all the evidence from other types of logs should be taken into account.

In deviated wells the calibration procedures outlined above are not entirely satisfactory. Clearly, in some structural situations it is not possible

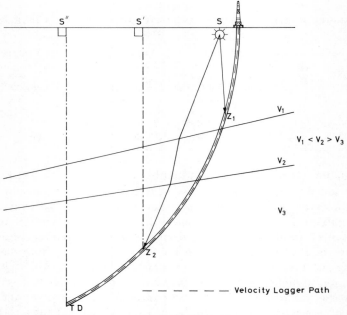

FIG. 7. Comparison of well geophone and acoustic log travel paths in a deviated well.

for the well geophone signal to propagate along the same earth paths as the velocity logger signals, as Fig. 7 illustrates, for a source position S close to the well.

At any point in the well, such as Z_1, Z_2, the velocity logger will have integrated all transit times observed along the path of the deviation which has a measured depth interval greater than the vertical distance both from the surface and between depth points. The well geophone signal may have followed a path totally outside the plane of the well, so that no similarity exists geometrically between the two sets of data. Not only are depth intervals not the same, so far as signal paths are concerned, the relative proportions of time spent by the signals in each layer are different. There is no unique solution to this problem and in some circumstances it may not even be possible to calibrate an acoustic log from well geophone data. In order to provide additional data with which to tackle the problem, further source positions such as S' and S'' in Fig. 7, chosen to be vertically above the well geophone, should be used in conjunction with the fixed position S for each level. Thus a set of near vertical and a set of slant times will be obtained for the entire range of geophone depths. By using the velocity log now as a fine earth layering model, and by rigorously computing the Snell's law refraction effect at each boundary, the slant geophone times may be adjusted to the vertical. The log itself, by using the directional survey results, may be reconstructed as though it were run in a vertical hole beneath the rig position, and reintegrated. Thus four sets of data now exist, one vertical and one slant for both well geophone and acoustic log, and comparison of these data will provide a better solution to the overall velocity distribution. It should be noted that the times recorded by the well geophone from source positions S' and S'' for example will most closely agree with the conventional surface seismic data, but that average velocities only between surface and detector can be determined with accuracy; subtracting the interval time between S' and Z_2 from the interval time between S'' and TD does not necessarily give the true vertical interval time between Z_2 and TD because different parts of the subsurface section are involved in each measurement.

There has been some change in recent years in the method of picking arrival times on well geophone records. Traditionally, a point was picked as early as possible in time near the onset of the first arrived signal, and was referred to as the first break. Assuming the earth to be a velocity dispersive and absorptive medium, which produces the pulse broadening effect, this point may represent the arrival time of a few specific components of frequency out of the total spectrum of frequencies being propagated. These times therefore may be characteristic of particular phase velocities and not the group velocity of the packet of energy with which the seismic method is concerned. Furthermore, the time at which a first break can be picked is a function of its amplitude and the recording gain of the instrumentation, and will vary from record to record, and from geophone level to geophone level. Ideally, the point on the waveform which characterises the group velocity, that is, the centre of the pulse, should be picked. This cannot readily be done in practice so some other easily recognisable and invariant point near the centre of the pulse is required. This could be, for example, the first zero crossing point of the geophone trace after the onset, but for practical purposes the first trough after the onset is now commonly used. This has the advantage of conforming with normal practice when mapping events on seismic cross-sections, where amplitude maxima or minima, not first arrivals, are used for correlation and time and velocity measurements. Thus the effects of absorption and dispersion are included in both sets of data.

In the calibration of acoustic logs, the trough method of picking geophone data usually minimises the scatter between time difference values, and will introduce into the calibrated log the absorption and dispersion influences which are inherent in the surface recorded seismic data.

The method of electronic picking of event times on acoustic logs uses neither first break nor maxima and minima points but something between them dependent on signal levels. Acoustic log signals are usually close to being mono frequency so dispersion is not a consideration, but being relatively high frequency (10–50 kHz), they will suffer absorption. This may not be significant for the spans and spacings currently employed, usually less than 3 ft, but may become so as these distances are increased.

VERTICAL SEISMIC PROFILING

As an aid to the better interpretation of surface recorded seismic data, and better understanding of the mode of propagation of both primary and multiple reflections, the conventional well geophone survey has been extended and modified in very recent years to provide considerably more data from each field survey. In this context the vertical seismic profile (VSP) is probably the most important single development in borehole geophysics for many years. Although used in the Soviet Union for about twenty years as an analytical method in various research programmes, only very recently has it been utilised in the West for purely commercial applications.

In its simplest form the VSP requires geophone check levels much more closely spaced (≤ 100 ft) than conventional velocity surveys, full wave train recording over several seconds of the geophone signal, and a fixed offset source capable of repeatable short duration signals. For every geophone position in the well, the detector will observe three main classes of event each quite distinct from the other:

(1) The first arrived signals, which have travelled direct from source to detector, together with any surface ghost reflections.
(2) Reflected events from interfaces beneath the detector, together with their multiples which travel upwards past the detector, involving an odd number of reflections.
(3) Downward travelling events reflected from the underside of interfaces above the well geophone, together with any multiples of these, involving an even number of reflections.

As the geophone is moved from one position to the next, the recorded time of each type of event will change in a characteristic manner as a function of the velocity distribution of the geologic column being examined. In Fig. 8 each trace is the full wave signal recorded at one geophone position in the well, and the data have been sorted sequentially in order of depth from top to bottom. Each trace has been time corrected to datum after adjustment for source offset and depth, and some gain compensation has been applied for the effects of spherical divergence and other amplitude losses.

In this display the first arrivals are thus aligned in accordance with the time–depth function for this particular seismic velocity distribution. The display is dominated by events such as those labelled M_1, M_2, and M_3 which are aligned in parallel with the first arrivals, and it can be easily shown that these must be events travelling downward past each geophone position at the same velocity as the first arrival signal. Because they are displaced in time from the first arrivals, they must have been reflected, at least twice, the last being from the underside of interfaces above the geophones and are the class (3) events as just described. Class (2) events, reflected upwards from beneath the geophone, will have an alignment equal and opposite to that of the first arrivals such as those labelled P_1, P_2, and P_3. The point at which these events intersect the alignment of the first arrivals defines the position in time and depth of the reflector.

In the general case, reflection coefficients at the surface and in the near subsurface may be expected to be greater than those at depth. As a consequence of this and the relative travel paths involved, the downgoing events will usually be of much larger amplitude than reflected events coming up from beneath the geophone. Furthermore the downgoing events will have traversed the near surface layers at least twice so that their pulse shape will be characterised by a progressively larger loss of high frequency than the upward reflected events. Both of these effects are apparent in Fig. 8.

In order to examine separately the nature and origin of upward and downward travelling events the data can be redisplayed so that both types of events appear as near horizontal alignments. In Fig. 9, each trace of Fig. 8 has been statically shifted by an amount equal to its first arrival time, so that the slope through the first arrivals is now double what it was. After the application of some standard seismic signal enhancement routines the upward travelling reflected events now appear clearly as near horizontal. The two-way reflection time and depth can be observed directly from the intersection points on the first arrival curve and can be correlated with the acoustic log plotted against a linear travel-time scale. Where the alignment

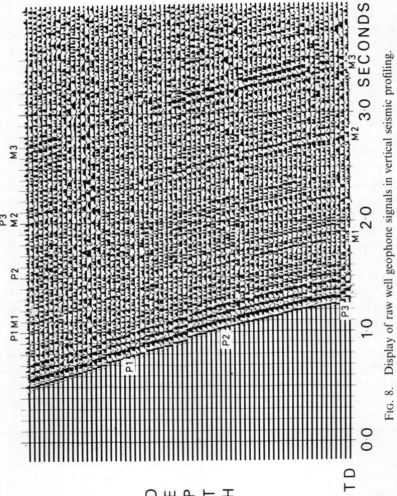

FIG. 8. Display of raw well geophone signals in vertical seismic profiling.

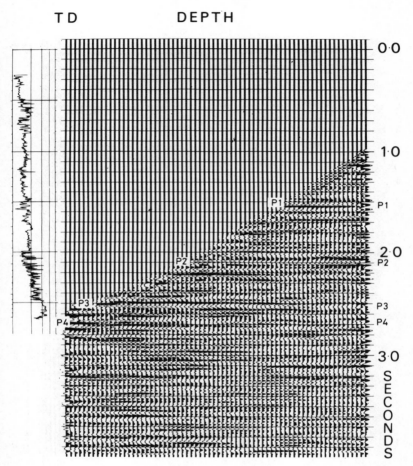

FIG. 9. Vertical seismic profile data showing enhanced primary reflections and upward travelling signals.

of the reflections departs from the horizontal, the deviation is a function of the angle of true dip of the reflector.

Figure 10 shows the data from Fig. 8 statically shifted in the opposite sense to that of Fig. 9, so that the first arrival times of each trace are reduced to zero time. The horizontal alignments indicate the first arrivals at zero time, and the downward travelling multiple reflections. The time sequence of these downward travelling multiples will be contained precisely

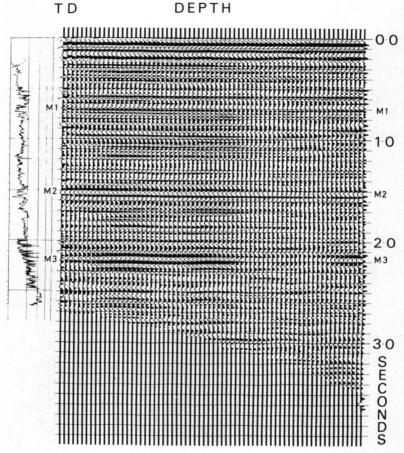

FIG. 10. Vertical seismic profile data showing enhanced downward travelling signals and multiple reflection pattern.

in the surface recorded conventional seismic record, because these same events will subsequently be reflected back to the surface from the same reflectors which generate the primary reflections. Except for scaling, the transfer function for the return path is the same as that travelling downwards, so that the waveform observed at the surface is the convolution of the downgoing wave with itself at the reflector, and will appear as the tail to all primary reflections on the seismic section.

OFFSET SOURCE SURVEYS

Because of the difficulties of calibrating velocity logs obtained in deviated wells (described in the section on calibration of velocity logs) and the need to obtain true vertical average velocities in many areas, the technique of running well geophone surveys in deviated wells has emerged as a standard field method. With the advent of VSP field and data processing techniques,

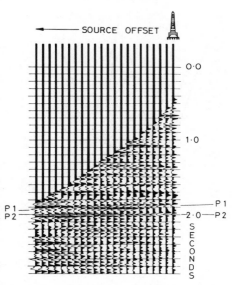

FIG. 11. Vertical seismic profile data from a deviated well showing primary reflection and upward travelling signals.

a further dimension has been added to deviated well velocity surveys. Figure 11 illustrates the type of seismic data that can be acquired from such well geophone observations. Each trace is the signal detected by the geophone at equally spaced intervals of depth along the path of a deviated well; shallowest geophone levels are on the right of the figure and deepest on the left. Each well geophone station was recorded with the source vertically above the geophone. Primary reflections can be seen throughout the section penetrated by the well, in addition to two separate events labelled P_1 and P_2 from beneath total depth of the well. These two events are sub-parallel and indicate a structural effect just above 2 s.

A variation of the VSP technique, in which a fixed source position is used

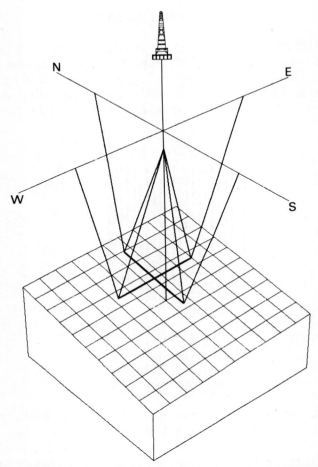

FIG. 12. Geometry of travel paths for an offset source velocity survey using a fixed well geophone location.

with the well geophone occupying successive equally spaced positions in the well, is to fix the geophone position and move the source (see Fig. 12). When the source is allowed to occupy equally spaced positions on two perpendicular lines through the well, two separate equivalent single cover seismic records are obtained, one of which is shown in Fig. 13. The well geophone now becomes a buried detector capable of mapping points on a reflector away from the well. The spread of these points laterally is a

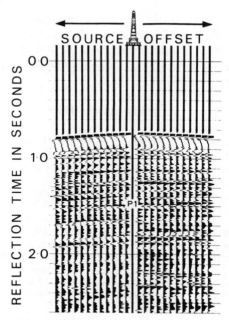

FIG. 13. Raw data from an offset source velocity survey using a single well geophone location.

function of the surface source offset distance and the distance of the geophone from the reflector being mapped. As with the standard VSP, both upgoing and downgoing events will be recorded.

One advantage of this technique is an improvement in the primary–multiple reflection amplitude ratio because the geophone can be placed well below the near surface reverberant systems and closer to the source of primary reflections. In addition, examination of the moveout effects both on the direct arrivals and subsequent events provides information on structure and lateral velocity variations.

FUTURE DEVELOPMENTS

In some notable exploration areas, present seismic techniques are approaching the limit of their capabilities to resolve some important structural and sedimentological problems. Borehole geophysical techniques may therefore be expected to aid and extend the range of surface to surface methods. Thus further developments may be anticipated in the design of energy sources, the type and multiplicity of down-hole detectors, data processing and analytical studies of the well geophone signals, and improved design and performance of continuous acoustic logging sondes.

BIBLIOGRAPHY

1. DIX, C. H., *Seismic prospecting for oil*, Harper's Geoscience Series, New York, NY, USA, 1952.
2. GUYOD, H. and SHANE, L. E., *Geophysical well logging*, Vol. 1, Hubert Guyod, Houston, Tex, USA, 1969.
3. GRANT, F. S. and WEST, G. F., *Interpretation theory in applied geophysics*, McGraw-Hill, New York, NY, USA, 1965.
4. GAL'PERIN, E. I., *Vertical seismic profiling*, SEG special publication No. 12 (ed. J. E. White), Box 3098, Tulsa, Okla, USA, 1974.
5. KENNETT, P. and IRESON, R. L., Recent development in well velocity surveys and the use of calibrated acoustic logs, *Geophys. Prospecting*, **19,** No. 3, p. 395, 1971.
6. KENNETT, P. and IRESON, R. L., Some techniques for the analysis of well geophone signals as an aid to the identification of hydrocarbon indicators in seismic processing, 1973 *Annual Meeting of Society of Exploration Geophysicists*, Mexico City, Mexico.
7. KENNETT, P. and IRESON, R. L., Vertical seismic profiling: Recent advances in techniques for data acquisition, processing and interpretation, 1977 *Annual Meeting of Society of Exploration Geophysicists*, Calgary, Canada.
8. MICHON, D., *Vertical seismic profile* (unpublished).
9. THOMAS, D. H., Seismic applications of sonic logs, *Fifth European Society of Professional Well Log Analysts' Logging Symposium*, Paris, France, 1977.

Chapter 4

SEISMIC SOURCES ON LAND

W. E. LERWILL

Seismograph Service (England) Limited, Keston, Kent, UK

SUMMARY

A basic mechanical model of a seismic vibrator illustrates the way in which it is intended to generate the compressional wave but it is clearly inadequate to describe the overall response of the system. Hence, the electrical analogy of a simple mechanical system is introduced in order to develop the analogue of a practical vibrator working into the complex load impedance of the earth. By this means, the amplitude and phase response of the baseplate when it is driven with a sinusoidal force is shown to be equivalent to voltage and current components in a simple network. A similar analogue illustrates the action of a weight drop source which is then compared with the vibrator and other energy sources.

INTRODUCTION

A brief history of seismic exploration by B. B. Weatherby appeared in the Society of Exploration Geophysicists (SEG) journal *Geophysics* in 1940, in which he describes a seismic experiment carried out by Robert Mallet in 1846. Mallet used gunpowder for his energy source which he fired electrically by closing a switch when he started his recorder. Again, in 1876, General H. L. Abbot 'took advantage' of an explosion of 25 tons of dynamite to measure the velocity of seismic waves; then he told Mallet that his estimates were wrong. So, on the face of it, there is not a great deal of change to report in seismic exploration practice for the past 130 years.

Chemical explosives have continued to be the most popular energy source on land since those early days. Dynamite is still one of the most convenient means of storing and transporting the enormous amounts of energy required for a seismic survey. But by far its most useful feature is the relative ease with which dynamite is planted deep in the earth, where its broad-band, high-energy pulse is most effective; for this reason alone we can be sure that it will continue to be used in the foreseeable future.

Nevertheless, there will always be a need for non-explosive sources where, for various reasons, it may be inconvenient to drill shot-holes, or where the use of dynamite is prohibited. However, the surface source suffers a two-fold handicap as follows:

(1) The energy must penetrate the highly attenuating layers in the near surface, which act as a severe high-cut filter on the spectrum of the transmitted pulse.

(2) Since it is on a free surface, the source is a very inefficient generator of the pressure wave, therefore it requires massive equipment to inject sufficient energy for deep penetration.

Much has been accomplished to overcome these problems over the years. The several types of impulse generators and vibrators working in the field today produce excellent results which are proof enough of their success. However, it is interesting to review the problems in point (2) concerning the efficiency of surface sources. The following pages contain a simple account of the way in which a vibrator baseplate responds to an input force when it is pressed against the earth's surface.

A mechanical model of the basic vibrator is given in order to develop its electrical analogue, from which it is possible to see the problem of injecting maximum power into the earth in terms of a simple network. The load presented by the earth indicates that, for a given baseplate area, there is a cut-off frequency below which the vibrator is a very inefficient radiator. The same applies to all surface sources whether they are swept frequency or impulse generators, since they all pass through the same earth filter. The description does not attempt to account for the way in which the total energy in the sweep, or the impulse, is distributed in the earth.

It will be assumed that the reader is familiar with seismic reflection survey practice, and that the Vibroseis® technique needs no explanation. The elementary electrical equations will be more familiar to those with some a.c. theory, but the main argument does not stray far from general knowledge.

® Trade mark of Continental Oil Co.

THE SEISMIC VIBRATOR

Most seismic vibrators in the field today are designed specifically to generate the compressional wave. Practically all of those employed in deep oil exploration use electrically controlled hydraulic systems to keep the baseplate motion in some fixed phase relationship with the swept frequency drive signal. These are important refinements but they will not be considered in this chapter, which is concerned only with the basic dynamics of the method.

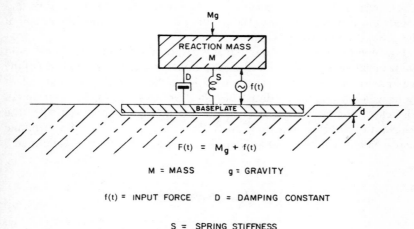

FIG. 1. Basic model of a vibrator.

The underlying principle of the vibrator is illustrated in Fig. 1. A heavy reaction mass is supported by a spring S with its damping factor D, both of which represent the mechanical impedance of a force generator $f(t)$. The weight of the reaction mass is shown as a positive force Mg acting against the baseplate. Therefore, when the system is at rest, the baseplate applies a static pressure on the ground resulting in a displacement d which, assuming the earth is elastic, will return to zero when the pressure is removed. The swept frequency signal is applied by various means which will be described later, but for the moment it may be considered as an oscillatory force acting between the baseplate and the reaction mass as indicted by $f(t)$ in the figure. Therefore, the equation for total force $F(t)$ acting on the baseplate is as follows:

$$F(t) = Mg + f(t) \qquad (1)$$

where M = mass and g = acceleration due to gravity. The force function $f(t)$, for the time being, may be considered as a sinusoidal drive signal of constant frequency ω which will produce a peak acceleration A on the mass M so that:

$$f(t) = MA \sin \omega t \tag{2}$$

Therefore, the total force is given by:

$$F(t) = Mg + MA \sin \omega t \tag{3}$$

Thus, over one complete cycle of ω, the input force swings positive and negative which in turn causes the required compression and rarefaction of the ground under the baseplate.

It is obvious from eqn. (3) that if A is greater than g the total force could swing negative or, in other words, if the input force exceeds the weight of the reaction mass the whole system will leap into the air during a negative half cycle, which would have a disastrous effect on the shape of the input signal. The effect that is aimed at is illustrated in Fig. 2; provided that the acceleration of the reaction mass does not exceed g, the dynamic displacement will be contained within that produced by the static load.

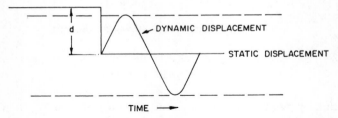

FIG. 2. Baseplate motion on elastic medium when a sinusoidal input force does not exceed g.

Note that the displacement in Figs. 1 and 2 is determined by the compliance of the ground. On very soft ground d will be large but on very hard rock surfaces it may well be too small to notice, but provided g is not exceeded there will be no distortion due to the baseplate leaving the ground. However, the peak force required from a vibrator designed for oil exploration is equivalent to 7000 kg or more, which is not considered a practical size for the reaction mass. Therefore, in current practice, the total weight of the vibrator vehicle is used to hold the baseplate in contact with the ground when the force is made to exceed that of the reaction mass; a

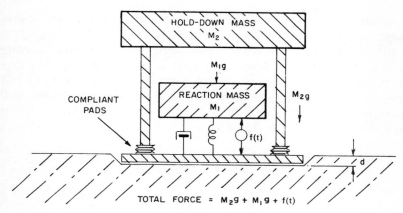

Fig. 3. Additional mass holding baseplate in contact with the ground.

model of this version is shown in Fig. 3. Exactly the same principle applies, but the total force is now given by the equation:

$$F(t) = (M_2 + M_1)g + MA \sin \omega t \qquad (4)$$

Thus, the input force can now equal the combined weights of M_2, the vehicle, and M_1, the reaction mass, before the system is in danger of leaping into the air. However, there are complications; it is not desirable to vibrate the vehicle, nor to transmit the noise from its generators into the baseplate, therefore compliant pads are fitted to act as vibration isolators as indicated in the figure; these are carefully chosen to provide effective isolation over the sweep spectrum, and at the same time avoid unwanted resonance.

THE ELECTRICAL ANALOGUE

There are many factors not indicated in Fig. 3 that also affect the response of the vibrator. The system that provides the drive force, if it is hydraulic, will contain flow pipes and control valves; or if it is electromagnetic, will have inductance in the moving coil and the response of the power amplifier, all of which must be taken into account. Not least of all is the response of the earth in the near surface, which would be difficult to indicate in a simple diagram, and the more of these we try to include in a mechanical model the more complicated it becomes.

It was in order to simplify the design of such systems that the electrical analogue was first introduced. It is a convenient way of setting out a complicated problems in terms of network equations for which there are well-known solutions. Although it is beyond the scope of this chapter to describe it rigorously, those who have had some experience with elementary a.c. theory may find this simplified description interesting, and they may also be surprised to see how revealing the analogue of the system in Fig. 1 turns out to be. However, it is important to know exactly what the electrical parameters are intended to mean, otherwise it will fail completely in its purpose to simplify the problem.

TABLE 1
MASS–CAPACITANCE ANALOGY

Mechanical			Electrical		
Mass	M	kilogrammes	Capacitance	C	farads
Compliance	$1/S$	metres/newton	Inductance	L	henries
Damping Constant	$1/D$	metres/second/newton	Conductance	$1/R$	siemens
			Voltage	e	volts
Particle Velocity	v	metres/second	Current	I_{dc}, i_{ac}	amps
Force	F	Ma newtons	Flux Linkage	ϕ	webers
Displacement	d	$\int v\,dt$ metres	Power	W	ei watts
Power	W	newtons metres/second	Electrostatic	E_k	$\frac{1}{2}Ce^2$ joules
Kinetic Energy	E_k	$\frac{1}{2}Mv^2$ joules	Magnetic	E_p	$\frac{1}{2}\phi^2 L$ joules
Potential Energy	E_p	$\frac{1}{2}Sd^2$ joules			

The traditional analogy is the one in which mass is represented by inductance, from which it follows that a mechanically resonant system can be conveniently represented by a series resonant circuit in which voltage is proportional to force. A more recent alternative represents mass by capacitance so that the same mechanical system becomes a parallel resonant circuit in which current is proportional to force. They are called the Mass–Inductance (M–L) and Mass–Capacitance (M–C) analogies.[1] The M–C convention was first introduced because it simplified the somewhat tricky problem of including the real electrical circuit in the analogue of a mechanical system, where it happens to be much more convenient to let current be proportional to force, and, since we will be considering electrodynamic systems, it will be used in this description; the equivalent parameters are listed in Table 1.

The table does not suggest that there is a similarity between current and force, for example, in the same way that current might be thought of as water flowing through a pipe; it is merely a device to put the dynamic equations of mechanics, or in this case the diagrams, into a more convenient form for solution. Figure 4 shows how it applies in a very simple case. The mechanical system in Fig. 4(a) represents a spring supporting a mass which is free to move in the vertical direction only, its electrical analogue, which is shown in Fig. 4(b), is the well known parallel resonant circuit. It will be assumed that the simple principles involved need no explanation, and it is understood that both systems will oscillate at their own natural frequencies when they are excited with a suitable impulse. The formulae for the

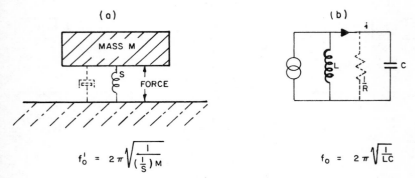

$$f_0^1 = 2\pi\sqrt{\frac{1}{(\frac{1}{S})M}} \qquad f_0 = 2\pi\sqrt{\frac{1}{LC}}$$

FIG. 4. Illustrating the electrical analogue (b) of a simple mass spring system (a).

undamped natural frequencies are shown in the diagram, the similarity between these is the essence of the analogy.

Damping losses that would occur in the spring may be accurately represented by a resistor in the electrical analogue; however, it should be noted that, since damping is represented by conductance in the M–C analogy, the resistor would appear in parallel with L and C in the circuit.

When a force is acting directly upon the mass it is analogous to a current introduced into the circuit. The constant current generator symbol which is used to indicate the input force in the diagram represents the odd concept of infinite source impedance. A series voltage generator, which would have zero source impedance, may be more familiar, but it would be appropriate here only if it was intended to illustrate a constant velocity input.

Before we leave the simple undamped circuit it will be useful to see how closely the parameters of force and velocity are related to current and voltage. The diagram in Fig. 5 is almost self-explanatory, but it is worth looking at in detail for a moment. Consider the mass in the figure oscillating at its natural frequency with constant amplitude; then the diagram shows the action in the mechanical system for one complete cycle at intervals of

$\pi/2$ rad. Starting at (a), using the convention where downward force on the mass is positive (as adopted in Fig. 1), the spring is exerting maximum positive force. This corresponds to a peak positive current in the analogue shown immediately below the mechanical system. In (b) the mass has reached its maximum velocity in its downward flight, which corresponds to maximum voltage across the capacitor. Thus the energy is said to be stored

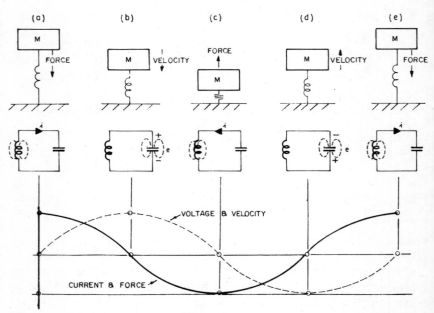

FIG. 5. Motion of mass spring system compared with voltage and current in the M–C analogy.

in the inductance when the mass is fully deflected, and in the capacitor when the velocity reaches its peak. The curves representing force and velocity (or current and voltage) are plotted at corresponding parts of the cycle at the bottom of the diagram. Notice that the phase relationship is consistent with force, which is directly proportional to acceleration, being the first differential of velocity, and that it also holds in the case of current and voltage. However, for the next step, which shows how a mechanical system is combined with an electrical one in the analogue, it is more useful to remember that the velocity of the mass represents a voltage across the parallel resonant circuit.

Figure 6(a) represents a moving coil drive unit coupled to the mechanical system similar to that in Fig. 5. The coil is supported in a cylindrical gap in which there is a magnetic field of strength B. When a current flows through the coil it generates a force f so that

$$f = Bli$$

where l is the length of wire in the coil and i is the current. The force acts so that the coil and the mass move together which, in turn, causes an emf to be

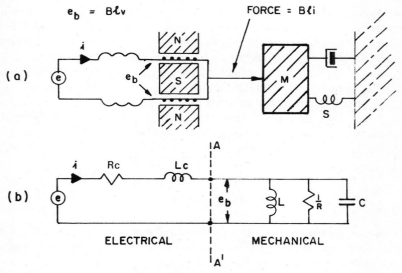

FIG. 6. Complete analogue (b) of electromechanical system (a).

induced into the coil in opposition to the supply voltage e; thus there is a back emf, e_b, in the electrical drive circuit where

$$e_b = Blv$$

in which v is the velocity of the mass.

The electrical part of the analogue in Fig. 6(b) contains the components e, R_c and L_c, which represent the alternating supply voltage, the d.c. resistance of the coil and its inductance respectively. In reality they are the only components in a continuous electrical circuit, but it would be correct to include in the diagram a second generator to represent e_b, the back emf induced into the coil. This is implicit in the following expression which describes the total voltage in the circuit:

$$e = j\omega L_c i + R_c i + e_b$$

where ω is the frequency of the supply and j is an operator. But, as already indicated, it is equally true to write:

$$e = j\omega L_c i + R_c i + Blv$$

in which the last term is the clue to a natural relationship between the electrical and mechanical circuits. Instead of inserting a generator to represent e_b, points A–A' are connected across the inductor and capacitor in the mechanical analogy where, as shown in Fig. 6(a), the voltage Blv appears. Thus a realistic link between the two circuits is established; current and voltage from the electrical circuit will now behave like force and velocity in the mechanical system. All that remains is to find the appropriate scaling factor that will relate the units of real mechanical components to those of the electrical components in the analogue.

Consider the mechanical part of the circuit which is shown on the right of points A–A' in Fig. 6(b). If the analogy is to be numerically correct then the inductance of L and the capacitance of C must be such that the following equation is true:

$$i = e_b \left(\frac{1}{j\omega L} + R + j\omega C \right) \quad (5)$$

where i is the electric current in the electrical part of the circuit and e_b is the true back emf. Now according to the M–C analogy the force into the mechanical part of the circuit could be described as follows:

$$Bli = \frac{e_b}{Bl} \left(\frac{1}{j\omega \left(\frac{1}{S}\right)} + D + j\omega M \right) \quad (6)$$

since force = Bli and velocity = e_b/Bl. Therefore, all that is necessary to put eqn. (6) into the required form is to divide it throughout by the constant, Bl, which gives:

$$i = e_b \left(\frac{1}{j\omega \frac{(Bl)^2}{S}} + \frac{D}{(Bl)^2} + j\omega \frac{M}{(Bl)^2} \right) \quad (7)$$

from which it is clear that the analogy is true when the components L, R and

C in eqn. (5) are derived from mechanical parameters which are scaled as follows:

$$L = \frac{(Bl)^2}{S}$$

$$\frac{1}{R} = \frac{(Bl)^2}{D}$$

$$C = \frac{M}{(Bl)^2}$$

When it is required to find true force or velocity from the current and voltage in the analogue, it should be remembered that:

$$\text{force} = Bli$$
$$\text{velocity} = e_b/Bl$$

In passing, it is interesting to note that the factor Bl is common to all such moving coil transducers. A notable example is the geophone in which, since it is basically a generator, the sensitivity, or the 'transductance' as it might be called, is given in volts per unit displacement per second. A typical figure for a present-day geophone would be about 1.5 V/cm/s, all of which is a reminder that the output from the geophone is directly proportional to the particle velocity.

Vibrator drive units are specified in terms of force per ampere (d.c.). One which is powerful enough for seismic use would deliver something in the order of 100 N (10 kg weight)/A.

THE ANALOGUE APPLIED

The simple example shown in Fig. 6 is almost identical with the basic vibrator in Fig. 1; the input force acts upon the mass in exactly the same way, therefore it is not difficult to imagine its analogue. However, the way in which the energy is transferred to the earth must also be represented by an output, and where it should appear in the network may not be immediately obvious from the mechanical diagram. The correct choice can be made by stating what is happening at the baseplate in terms of force and velocity in the model, and then comparing that with what it means in terms of its analogue (see Table 2).

Both conditions are satisfied when the earth load is connected in series

TABLE 2
EFFECT OF EARTH LOAD IMPEDANCE

Mechanical diagram	Electrical analogue
Force applied equally to the reaction mass and the earth via the baseplate.	Current flows equally through the capacitor and the earth load.
Maximum spring displacement is shared between the reaction mass and the earth via the baseplate.	Total potential energy in the inductance, therefore total voltage is divided between capacitor and earth load.

with the capacitor at points A and B in Fig. 7. The next step is to choose a network to represent the earth load, or at least, since it would be impossible to simulate exactly the conditions met in the field, to choose one that accounts for the fact that a compressional wave is transmitted.

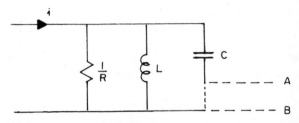

FIG. 7. Analogue of basic vibrator in which the effect of the earth load appears between points A and B.

In general, then, the earth behaves like a heavily damped compliant system in which the energy from the vibrator is radiated and absorbed. Quite a large proportion must be spent in compacting and heating the soil immediately beneath the baseplate, but for the moment we will consider how efficiently it is delivered rather than where it goes. Therefore, to give some basis for discussion, it will be represented by the simple network in Fig. 8, where the inductance L_g behaves like the compliance and the conductance $1/R_g$ will account for the total energy that is dissipated in the ground.

The network responds in much the same way we would expect a vibrator baseplate to behave when it is pressed against an elastic earth, as Table 3 illustrates.

Notice that Z_g in Fig. 8 is the admittance of the real earth, since

TABLE 3
RESPONSE OF SIMPLE LOAD NETWORK

Real behaviour	Analogous behaviour
A force due to the weight of the reaction mass is applied to the baseplate.	A d.c. current is applied to the network.
Baseplate is pressed into compliant earth until it reaches equilibrium.	Voltage e_B appears across network then decays exponentially as the inductance L_g is charged, until e_B is virtually zero.
When weight is removed the baseplate returns to original position.	Negative voltage $-e_B$ appears across network as L_g discharges its energy into $1/R_g$.

compliance and the reciprocal of the damping factor are used in the M–C analogy. Furthermore, there are real and imaginary terms shown in Z_g when it might be customary to think of the earth's 'acoustic impedance' in the reflection coefficient as being purely resistive. The full explanation for this will be found in the theory of acoustic wave propagation[2] but, since it is important that we agree on the definition of impedance, it deserves a brief comment here. What is commonly accepted as 'mechanical impedance' is the ratio of force and velocity, thus:

$$|Z_M| = \frac{\text{Force}}{\text{Velocity}}$$

and since force and velocity are not necessarily in phase, the impedance is complex; this comes from the M–L analogy.

Now acoustic impedance to a compressional wave is based on the same concept except that force becomes pressure, and therefore:

$$|Z_A| = \frac{\text{Pressure}}{\text{Particle velocity}}$$

where it is also true that, in the general case, Z_A is complex. However, when a plane wave is travelling through the body of the earth, pressure and velocity are in phase, therefore Z_A is real for all frequencies. These conditions are substantially the same for the spherical wave in the far field, which is taken to be the case in seismic reflection survey. But it is worth remembering that it is only under this plane wave assumption that the familiar equation is true in which

$$Z_A = \frac{P}{v} = \rho c$$

where v = particle velocity, P = pressure, ρ = density and c = wave velocity.

There is an elegant analogy, which is discussed by Kinsler and Frey,[2] between the acoustic impedance to a plane wave and the characteristic impedance of a transmission line. A transmission line contains inductance and capacitance distributed along its length, but when they are uniform, or in other words when a coaxial line is perfectly concentric along an infinite length, it has an impedance which is given by:

$$Z_0 = \sqrt{\frac{L}{C}}$$

where Z_0 is resistive at all frequencies.

Now it would not be practical to use a baseplate large enough to simulate the plane wave at seismic frequencies, therefore our load is not purely resistive, which is why the L_g component appears in Fig. 8. If we stretch the analogy to its limit, it is as if we are looking into the reactive part of the earth's transmission line which at low frequency (in the M–C analogy) is inductive. The vector diagram in Fig. 8(b) shows how the phase of the

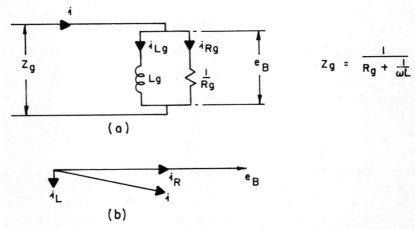

FIG. 8. (a) Simple analogue of the load presented by the earth. (b) Phase relationship between voltage and current components in the network.

currents are related in the network; this will be referred to later on in connection with its response.

The electrical analogy of a vibrator complete with its earth load is shown in Fig. 9(a). This consists of the simple mass–spring system of Fig. 1 driven by the moving coil unit of Fig. 7. It will be appreciated that a more complete analogue would include the mass of the moving coil, and other components that might be significant. The mass of the baseplate is represented by the

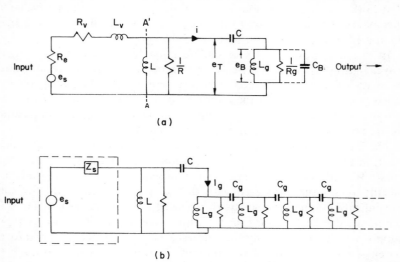

FIG. 9. (a) Complete analogue of working vibrator. (b) Equivalent analogue showing earth load as a transmission line.

capacitor C_B simply to show where it would appear, but since its reactance is very high compared with that of L_g it will not be significant at normal sweep frequencies.

The components to the left of the line A–A' are those in the electrical part of the moving coil unit. L_v and R_v are the inductance and resistance of the coil which is driven by a primary power supply e_s with its internal resistance R_e. All of these will play a vital part in the overall response of the system, but the details need not concern us here. The important point they illustrate is that the drive system, in common with any other, has an effective source impedance which must be taken into account when it is expected to deliver maximum power into a load.

The most interesting feature is the way in which the main L and C components would naturally link in with those of the earth's transmission line, if the baseplate could be made large enough to obtain a perfect impedance match. This is illustrated in Fig. 9(b) where the L_g and C_g components represent the compliance and mass equivalent of 'lumped constants' in the earth. Under these conditions it is clear that there would be an optimum L–C ratio for the compliance and reaction mass in the vibrator. All that would be necessary to obtain a perfectly flat frequency response would be to ensure that the source impedance, Z_s, was equal to that of the earth. Furthermore, 50 % of the power from the source would then be spent entirely in generating the pressure wave (the other 50 % being dissipated in Z_s).

In practice, however, the baseplate must be small compared to the wavelength of the wave it is expected to generate; consequently it behaves like a rather poor point source from which only about 8 % of the power it receives is transmitted in the compressional wave.[3,4] Consequently it is driven with as much force as possible, regardless of the earth's impedance, in the hope that some useful power is radiated.

The only criterion for choosing the suspension spring, or its equivalent, i.e. L in the analogue, is that it should not be resonant with the reaction mass, M, within the sweep spectrum. Consequently the reactance of L is usually very large compared with that of C, so that, except for some magnification at low frequency, the effect of L can be neglected. Thus, for the moment, we can regard the vibrator as consisting of a primary power supply which, by some means, drives a constant force sweep into the baseplate; or, in terms of the analogy, a constant current through the capacitor C into the network representing the earth.

Next we will consider the way in which the baseplate responds to the constant force. This can only be approximate with such a simplified earth load, but it does explain the type of performance that will be familiar to those who have worked with vibrators or the records they produce. Only the low frequency end of the spectrum will be considered in the following, since the capacitive reactance is not in the earth network. The baseplate velocity is represented in Fig. 9(a) by the voltage e_B which is given by the following equation:

$$e_B = \frac{i}{R_g - j\left(\dfrac{1}{\omega L}\right)} \qquad (8)$$

where i is the constant current and ω is the signal frequency in rad/s. Now it will be clear from eqn. (8) that the baseplate velocity will not be constant at all frequencies; when ω is small e_B tends to zero, and when ω is large e_B is equal to $i(1/R_g)$. In fact it behaves like a low cut filter with a 6 dB/octave slope. The curve in Fig. 10 shows e_B plotted against frequency for the purely arbitrary case where $1/R_g$ is equal to ωL when the frequency is 5 Hz.

It is technically difficult to maintain a constant input force at very low frequency, therefore it might be impossible in practice to determine a cut-off as clearly defined as that in Fig. 10. Nevertheless, when the baseplate is

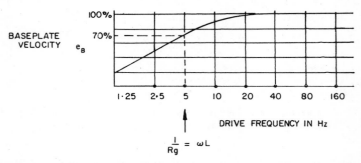

FIG. 10. Baseplate response at low frequency when drive force is constant.

acting as the radiator, it is clear that its performance will fall off at low frequency. Furthermore, it will be found that the point at which $1/R_g$ is equal to ωL will appear further down the spectrum as the baseplate area increases. This is an effect which, although it might not be described in the same terms, is shown to occur in acoustic radiators, loudspeakers and radio aerials.[2,5]

However, the Fig. 10 curve should not be taken too literally since there must be times, particularly when vibrating on a concrete road surface, when it would be difficult to determine the area of the effective radiator. Under these circumstances the vibrator, and the truck, behave as a shaker rather than the transducer we are considering here.

Regarding its phase response, the baseplate cannot be expected to follow exactly the same phase as the input force, as the following description shows.

The total current in the network can be found from eqn. (8) as follows:

$$i = e_B \left(R_g - j\left(\frac{1}{\omega L}\right) \right) \quad (9)$$

which may also be written:

$$i = i_{R_g} - j i_{L_g}$$

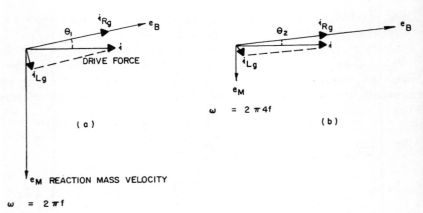

Fig. 11. Phase relationship between the reaction mass velocity, baseplate velocity and drive force (a) at frequency f and (b) at frequency $4f$.

where i_{R_g} and i_{L_g} are the currents in the resistive and inductive branches respectively. This is plotted as a vector diagram in Fig. 11(a) where the constant current **i** acts as the zero phase reference representing the drive force. Notice that, since it is a parallel network, the voltage $\mathbf{e_B}$ is always in phase with the resistive leg, and that it leads the drive force by the angle θ where:

$$\theta = \tan^{-1} \frac{i_{L_g}}{i_{R_g}}$$

or, from eqn. (9):

$$\theta = \tan^{-1} \frac{\left(\dfrac{1}{R_g}\right)}{\omega L_g}$$

It follows that θ tends towards 90° as the drive frequency approaches zero, and also that the baseplate velocity leads the drive force by 45° when $(1/R_g)$ is equal to ωL.

The phase and amplitude relationship that could exist at the start and finish of a two octave sweep is illustrated in Figs. 11(a) and (b) respectively. Notice that the velocity of the reaction mass, represented by the vector $\mathbf{e_M}$, lags 90° behind the drive force, and that its amplitude diminishes with frequency. This is a familiar sight in a working vibrator, which is to be expected in the analogue because the drive current passes through the capacitive reactance of the mass as follows:

$$e_M = i\left(-\frac{j}{\omega C}\right)$$

Once again, the phase that is measured in practice will depend upon the area of the effective radiator, but it is interesting to see that the analogue provides a basis for interpreting phase measurements when a vibrator (which is not phase compensated) is behaving as a transducer.

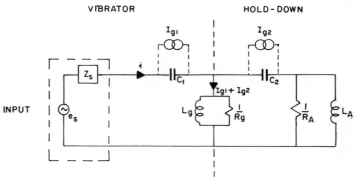

FIG. 12. Analogue of vibrator with hold-down weight.

Throughout this description it has been assumed that the input force does not exceed the weight of the reaction mass. However, in practice, we have the situation illustrated in Fig. 3 where the weight of the vehicle holds the baseplate in contact with the ground while the reaction mass acceleration exceeds that of gravity; the analogue of this arrangement is shown in Fig. 12. The reaction mass and the mass of the vehicle are represented by capacitors C_1 and C_2 respectively, and the d.c. current generators represent the static force that is applied to the baseplate by the weight of their masses. Therefore, the sum of the two d.c. currents passing through the earth load represents the maximum peak drive current i that can be applied to the network. The inductance L_A and conductance $1/R_A$ represent the compliant pads which isolate the vehicle from the baseplate.

Clearly, then, since the hold-down system appears across the earth load it must be designed to present a very high impedance to the drive current. It follows that resonance between L_A and C_2 must not occur anywhere near the sweep frequency, and the damping losses in L_A must be as small as possible (i.e. $1/R_A$ is large).

Regarding the effect of amplitude and phase distortion on the shape of the seismic pulse, it is a curious fact that a natural concern to preserve the form of the cross-correlation pulse in Vibroseis, has made us more acutely aware of the phase shifts that can exist throughout the entire seismic system. Also, as has been demonstrated above, the method itself provides a much more convenient means of measuring amplitude and phase response than the conventional impulsive method. Nevertheless, everything that has been

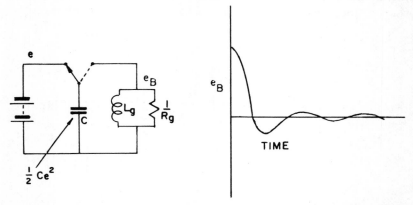

FIG. 13. Analogue of weight-drop source showing impulse response.

said above concerning the baseplate response must apply equally to any surface source with the same effective baseplate area; they must all look into the same highly reactive earth and respond accordingly.

Consider, for example, the action of a simple weight-drop source. Potential energy is stored in the mass when it is poised ready for the drop. When it is released it gathers momentum as it falls until, at the instant before contact with the ground, its kinetic energy E_K is given by the well-known formula:

$$E_K = \tfrac{1}{2}Mv^2$$

where M is the mass of the falling weight and v is its velocity immediately before impact. What happens next can be described in terms of the simple analogue in Fig. 13.

The capacitor is charged from a supply voltage e which represents the velocity of the mass immediately before impact, thus the kinetic energy is represented by $\tfrac{1}{2}Ce^2$. The impact is simulated by discharging the capacitor into the earth load as indicated by the switch. Thus the voltage e_B which appears across the network represents the velocity of the mass when it is in contact with the ground; in other words, the baseplate is now the surface area under the dropped weight. The characteristic response of the network is shown on the right of the diagram; there would be distortion due to bounce and many other factors in the real case, but this serves very well to illustrate the following point.

It would be impossible to say what force or velocity the falling weight delivers upon impact because the rate at which it loses momentum is unknown. Similarly, it is not possible to say what the current or voltage will be when a capacitor is discharged into an unknown load. In fact there is an analogy in the charge stored by a capacitor which can be seen in Table 4.

TABLE 4
MOMENTUM AND FORCE IN THE M–C ANALOGY

Mechanical	Electrical
Momentum $= Mv$	Charge $= Ce = Q$ (coulombs)
Impact force $= f = \dfrac{M\,dv}{dt}$ (newtons)	Current $= i = \dfrac{dQ}{dt}$ (amps)

In passing, it is interesting to note that although we have units of coulombs for a 'quantity' of charge, there is no equivalent unit in dynamics for momentum; however, the important points illustrated by the analogy are as follows:

(1) The network is excited by a charge, which is in effect a 'package' of energy with a potentially infinitely broad bandwidth, that is to say, a spike of energy not a step in current or voltage.
(2) The response it produces is determined by C as well as L_g and $1/R_g$ (or any other components in the real earth not shown in the analogy).
(3) The total energy in the charge is distributed in the wavelet it produces; thus a resonant system will rob energy from the rest of the spectrum.

Similarly, then, when the weight is dropped it delivers a package of energy which is, in effect, an infinitely broad-band spike. But the shape of the transmitted wavelet is determined entirely by the response of the earth, which must include the weight itself and the truck from which it is dropped.

The same filtering process is applied to any type of surface source whatever the shape of its energy pulse; it will be relentlessly convolved with the impulse response of the earth at the surface.

In the Vibroseis method, for example, the effective output from the vibrator can be considered as an autocorrelation pulse containing the total energy in the swept frequency. Therefore, if the sweep has sufficiently broad bandwidth, its response at the baseplate is identical with that from the impulse. The effect of phase shift can be removed by cross-correlating the geophone signal with the sweep that appears at the baseplate, or by phase compensation, but the amplitude response of the earth remains unaffected.

In practice it will be found that the resonant system which is excited by the impulse need not be entirely due to the mass of the source and the compliance of the earth. An example of natural resonance in the very near surface is illustrated in Fig. 14, where, in this case, dynamite was the energy source. The records show the result of a field experiment where the recording amplifiers were set at fixed gain with no automatic gain control

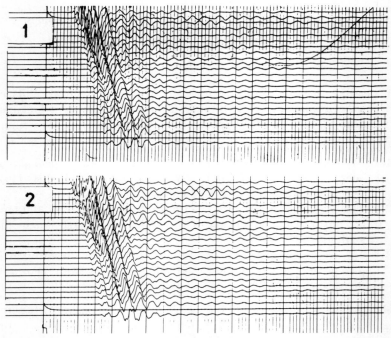

FIG. 14. Field monitor records illustrating the response of a natural resonant system in the near surface when the depth of the constant $\frac{1}{2}$ kg charge is (1) = 1·5 m, (2) = 3·0 m, (3) = 4·5 m, (4) = 5·0 m.

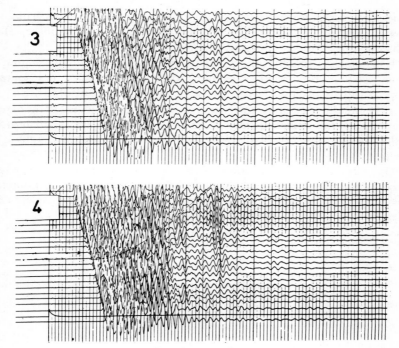

FIG. 14—contd.

(AGC) throughout the test, while 0·5 kg charges were fired at increasing depths at the same shot-point location. Records 1 and 2, which were taken when the charge was at 1·5 and 3 m, show a pronounced 30 Hz resonance with very little reflected energy. Records 3 and 4 show a dramatic improvement when the source is placed beneath the resonant system. In this area, a hard limestone plate was subsequently located in the clay at a depth of 5 m or so.[6] This illustrates a problem which faces all types of surface energy source alike, where dynamite has the clear advantage.

ENERGY

The energy available from the weight-drop unit is easily calculated from the potential energy in the mass of the poised weight, which, by the principle of conservation, is its kinetic energy upon impact. Thus:

$$E_\mathrm{P} = E_\mathrm{K} = Mgd = \tfrac{1}{2}Mv^2$$

where g is acceleration due to gravity. A vibrator, on the other hand, is usually specified in the peak sinusoidal force it delivers to the baseplate, which, on its own, says nothing about its power output.

It is not surprising that there is some difficulty in estimating the power from such a broad-band transducer when both its load and source impedances are reactive. Nevertheless, the same problem would arise if we were to attempt to estimate the instantaneous power dissipated in the earth from the falling weight, or any other source; the only way the vibrator differs, in this respect, is that it is not quite so easy to calculate its potential energy.

One method that has been suggested is to sum the square of the force over the duration of the sweep, and thus find what is in effect the momentum in the autocorrelation pulse, but it is not equivalent to the potential energy in the vibrator. This can be shown as follows;
Let
$$\text{force} = f = Ma$$
then
$$\text{momentum} = \int f \, dt = Mv$$
but
$$E = \tfrac{1}{2} Mv^2$$
therefore it follows that:
$$E_K = \frac{1}{2} \frac{[\int f \, dt]^2}{M} \tag{10}$$
or
$$E_K = \frac{1}{2} \frac{[Mv]^2}{M}$$

Thus, eqn. (10) suggests that the vibrator with a large reaction mass does not require as much energy in order to deliver the same momentum as a vibrator with a small mass; or what is more to the point here, it illustrates that the sum of the force does not properly describe the energy in the sweep.

Another factor that must be considered is that the force is not necessarily constant over the full range of the sweep, as it might be assumed to be in a formula derived from eqn. (10). It seems likely that it would be more reliable for the user to refer to the input power, which is usually specified, then compare that with the performance of other sources within the same part of the frequency spectrum. This is the method used with similar transducers or transmitters where there is the same kind of problem with the load. It would then be left to the manufacturer to provide response curves or figures for efficiency if it became possible to do so. The following examples show how the energy in the various sources compares.

Weight-Drop Source
A typical weight-drop unit drops a 3000 kg mass from a height of 3 m. Therefore, its potential energy is as follows:

$$E_p = Mgd = 3 \times 10^3 \times 9{\cdot}8 \times 3 = 88\,200\,\text{J}$$

where $g = 9{\cdot}8\,\text{m/s}^2$. This is considered to be a low-energy source which, for deep exploration for oil, delivers about 30 drops on one shot-point location; therefore the total energy which goes to make one record is $2{\cdot}6 \times 10^6\,\text{J}$.

The weight-drop source has the advantage of being simple and yet capable of remarkably deep penetration. This seems to be due to the fact that most of its energy is concentrated in the low frequency pulse, which is a characteristic impulse response in most areas.

The Vibrator
The input power to the hydraulic pump in the various types of vibrator ranges from 42 kW (56 hp) to 156 kW (209 hp). Thus, the energy delivered to a unit during a 14 s period, which is a typical length of sweep, can be expected to be between $0{\cdot}6 \times 10^6$ and $2{\cdot}2 \times 10^6\,\text{J}$; which, as a matter of interest, can be compared with the electromagnetic vibrators used in the early days of Vibroseis. They would have consumed between 60×10^3 to $140 \times 10^3\,\text{J}$.

The problem that has confronted the manufacturers in their effort to improve low frequency response is reflected in the way in which the units have grown over the years, both in power consumption and weight. They now match other sources in deep exploration with the advantage that they are able to work in built-up areas where the relatively small power output causes least damage. They also remain the only type of source with perfect control over the spectrum of the seismic pulse, which, apart from potential advantages in high resolution exploration, makes them peculiarly well suited for digital recording.

Since there is no significant source generated noise beyond the limits of the sweep, it is no longer necessary to employ the usual 72 dB/octave filter at $\frac{1}{2}$ Nyquist frequency in the recording instruments. Thus a sweep with a 125 Hz upper frequency limit may be sampled at 4 ms without there being a danger of aliasing coherent noise. This represents an important saving in the cost of recording and processing digital data.

Portable Rammer

The small rammer used in the Mini Sosie® technique demonstrates that a source with a small baseplate area is less inclined to excite ground roll. Some excellent high resolution results have been achieved in the near surface down to 500 ms or so, two-way travel-time. The maximum energy from the rammer is about 135 J; it might take between 200 and 1000 impacts to make one record, thus the total energy for this near surface source is between 27×10^3 and 135×10^3 J.

Dynamite

The energy stored in 1 kg of dynamite is estimated to be about 5×10^6 J. However, its efficiency as a source depends entirely upon the way it is used; the situation is similar to that of the weight-drop, where the energy is contained in what is virtually a broad-band spike but the wavelet it transmits depends entirely upon the reaction of the surrounding material. It appears to be very inefficient as a surface source when compared with the thumper, but that is to be expected when there is no overlying mass for its energy to react against, unless, like General Abbot, we use a charge of 25 tons.

ACKNOWLEDGEMENTS

The author wishes to thank the Directors of Seismograph Service (England) Ltd for permission to publish this contribution, without associating them with any of the opinions or conclusions that are reached.

Thanks are also due to his colleagues at Seismograph Service (England) Ltd, particularly to Mr J. J. Breugelmans for his advice, to Mr J. H. Tapping for making the diagrams, and to Mrs J. H. Plaister for typing the manuscript.

REFERENCES

1. BARKER, J. R., *Mechanical and electrical vibrations*, p. 16, Methuen & Co. Ltd, London, England, 1962.

® Trade Mark of Seismic Geocode Ltd.

2. KINSLER, L. E. and FREY, A. R., *Fundamentals of acoustics*, p. 122 section 5.8, p. 247, John Wiley & Sons Inc., New York, NY, USA, 1962.
3. MILLER, G. F. and PURSEY, H., The field and radiation impedance of mechanical radiators on the free surface of a semi-infinite isotropic solid, *Proc. Roy. Soc. (London), Ser. A*, **223**, p. 539, 1954.
4. MILLER, G. F. and PURSEY, H., On the partition of energy between elastic waves in a semi-infinite solid, *Proc. Roy. Soc. (London), Ser. A*, **223**, p. 68, 1955.
5. GAYFORD, M. L., *Acoustical techniques and transducers*, p. 77, MacDonald & Evans Ltd, London, England, 1961.
6. ZIOLKOWSKI, A. and LERWILL, W. E., *A simple approach to high resolution seismic profiling for coal*, p. 13, preprint of paper presented to EAEG meeting, Zagreb, Yugoslavia, 1977.

Chapter 5

MARINE SEISMIC SOURCES

R. LUGG

Seismograph Service (England) Limited, Keston, Kent, UK

SUMMARY

Over the last twenty years the marine seismic source has evolved from a single solid chemical charge, detonated as a high explosive point source, radiating near-field destructive energy in all directions, to an areal array of low energy mechanical sources, harmless to the environment but focusing its energy in the far-field. The early attempts to control dynamite and the gradual developments in a wide variety of low energy sources emerging to replace it are described.

Comparisons of intrinsic energy, period and band limited peak pressure are made. The many influential factors affecting the radiated far-field signature are discussed and illustrated. A general assessment of sources is given. Present day technology concerned with the control and measurement of the signature radiated by low energy arrays is described and illustrated.

INTRODUCTION

The seismic reflection method is an echo-ranging technique which penetrates the earth seeking to register and display its layers as continuous reflection alignments or correlating reflection sequences, in the form of a reflection time cross-section, from which the features of the geology of the true earth can be interpreted to predict structure. Ideally, a broad-band seismic source–receiver system would provide penetration to basement rocks with resolution of the closest spaced layers throughout the stratigraphic column. The ideal source would repeatedly radiate a spike signature whose unfiltered and unmodified shape would pass through the ideal earth, being reflected from discrete layers, and would be recorded without loss of bandwidth to be displayed as a perfect seismic time cross-section (for which the processing applies corrections related only to the source–receiver geometry).

In practice, however closely the source signature approaches a spike, it is broadened by absorption as it travels through the earth and lengthened by multiple reflections to become the propagating signature or wavelet which is neither short nor constant. Modern digital processing data enhancement techniques seek to extract the basic wavelet (assuming constant earth absorption) or the propagating wavelet, as steps in the prediction of the earth's response to an ideal spike. Whatever enhancement is achieved, resolution will be limited by the bandwidth of the propagating wavelet and will decrease with depth.

It is desirable that the source radiates a directional repeatable signature which either approaches a spike, or has a shape such that the bandwidth and energy level of the propagating wavelet associated with it, provide sufficient resolution at the depth of interest.

THE DYNAMITE ERA

The source used in early marine seismic exploration off the shore of California in 1948 consisted of a charge of dynamite of 80 lb weight fired at shallow water depth. Because of the high fish kill, and in response to political pressure from sporting and commercial fishing interests, the state permit governing seismic operations was revoked in 1949. Investigations undertaken by the Californian Fish and Game Commission show that the species of fish killed by explosions are mainly those with air bladders, which suffer an outward rupture due to the steep fronted negative pressure wave reflected from the surface of the water. Most of the more valuable food and game species of fish have such bladders.

In 1956, studies to provide a substitute explosive, reported by Jakosky and Jakosky,[1] resulted in black powder charges up to a maximum of 90 lb being allowed. These were equivalent to 9 lb of the previously used dynamite, and for a negligible level of fish kill achieved less penetration at approximately ten times the cost.

The pressure signatures of 1 lb dynamite and 45 lb black powder are compared in Figs. 1(a) and (b). The slower rise-time and lower peak

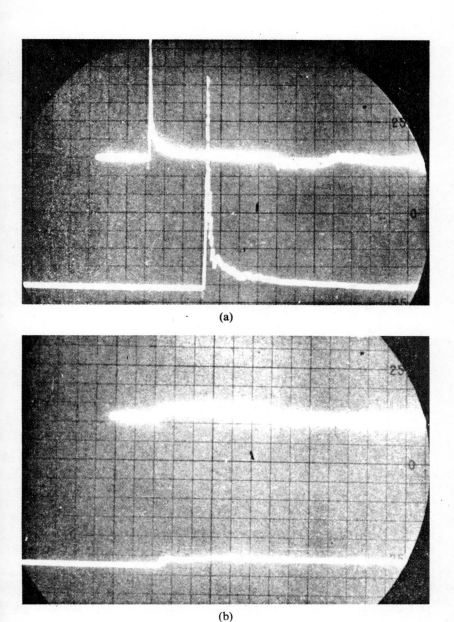

FIG. 1. (a) Pressure curves from a 1 lb charge of 40 % dynamite at a range of 25 ft. (b) Pressure curves from 45 lb of black powder at a range of 50 ft. Horizontal scales: 1 ms and 0·2 ms per division. Vertical scale: 38 psi per division. From Jakosky and Jakosky.[1]

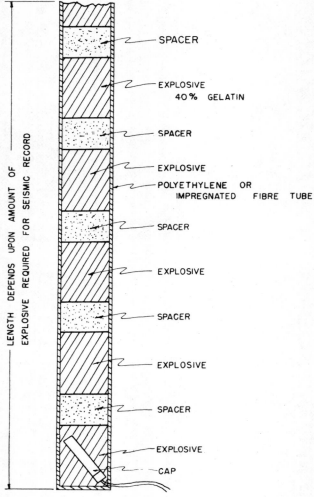

FIG. 2. Multipulse charge. From Jakosky and Jakosky.[1]

pressure of the black powder decrease its lethal properties and produce a higher percentage of low frequency energy, but produce a longer pulse of less resolving power.

Other studies by Jakosky and Jakosky to control the rate of the explosive reaction and reduce the overall velocity of detonation show that multipulse, a charge made up of a series of pellets of dynamite which fire progressively,

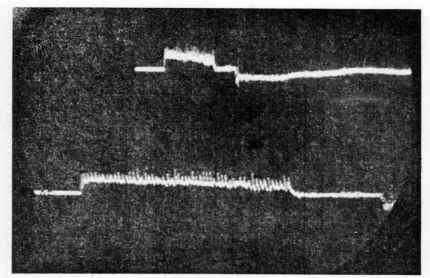

FIG. 3. Pressure curves from a 10 lb multipulse charge consisting of 50 pellets of gelatin. Range 100 ft. Scales as in Fig. 1. From Jakosky and Jakosky.[1]

provides a good compromise between penetration and fish kill. The construction of the multipulse charge is shown in Fig. 2[1] and its pressure signature is shown in Fig. 3. A comparison of explosive velocities for various materials is given in Fig. 4. Computed pressure time curves for 10 lb multipulse, 10 lb dynamite and 90 lb black powder are shown in Fig. 5.

The restrictions imposed in California did not apply to other off-shore areas where the normal practice, until well into the 1960s, was to fire 50 lb charges of nitro-ammonium nitrate 6 ft below the sea surface at intervals of 660 ft to obtain 6-fold common depth point coverage. Often 15 tons of this type of explosive (Seismex®, prepared as in Fig. 6) were expended in one day. This high explosive point source venting its gases into the air gave the shortest pulse of highest peak pressure and broadest bandwidth from the smallest, though expendable, package of energy capable of accurate repetitive firing.

It was accepted that approximately one third of its energy was used to lift the water spout (Fig. 7) which in falling back onto the water surface created

® Trade name of Imperial Chemical Industries Ltd.

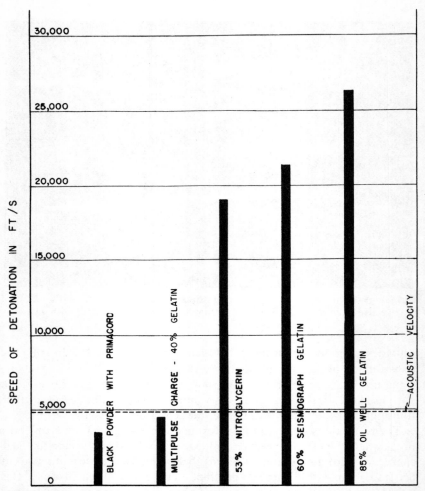

FIG. 4. A comparison of explosive velocities for various materials. From Jakosky and Jakosky.[1]

unwanted noise at later record times, that its very narrow pulse gave rise to high frequencies above the useful seismic band, and that this wasted energy was lethal to fish with bladders.

Sometimes these charges broke away from their firing lines to become a hazard if washed up on beaches or caught in fishing nets, and means were found to disarm them after a given time in the water. However, the main

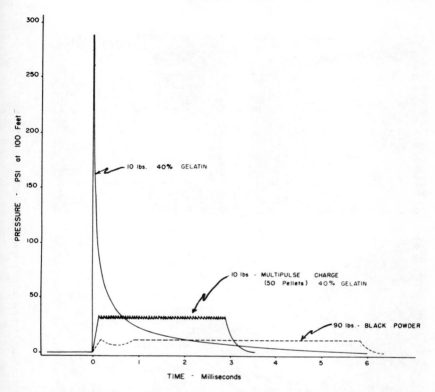

FIG. 5. Pressure time curves for 10 lb 40% gelatin, 10 lb multipulse and 90 lb black powder computed to the same base. From Jakosky and Jakosky.[1]

reason for most governments first limiting their size and then in 1969 prohibiting their use near the surface, was the environmental one of high fish kill.

By this time, experiments by Lavergne[2] and others had shown that marine seismic reflection amplitudes increase with charge depth to the extent that 30 times less dynamite is required at 30 ft water depth than at a shallow blow-out depth, to produce the same level of seismic response. Furthermore, they had shown that fish kill is reduced to negligible proportions, not only due to lower peak pressures from very much smaller charges, but also due to the greatly reduced negative peak pressures resulting from deeper charges.

However, the gain in seismic efficiency with charge depth, due to less

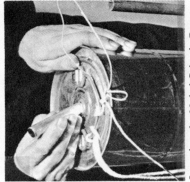
Inserting the Nobel Seismic Booster

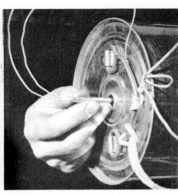
Inserting the detonator

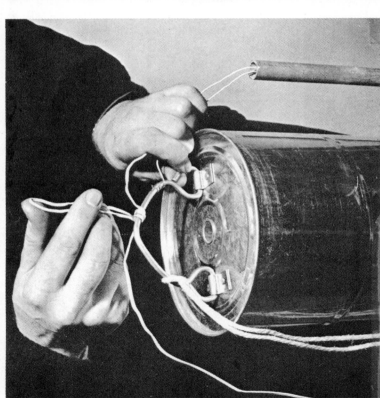
Attaching the leading wires

Fig. 6. Priming a 50 lb marine Seismex® charge. ® Imperial Chemical Industries Ltd.

FIG. 7. 50 lb seismic shot in the North Sea. From Imperial Chemical Industries Ltd.

erosion of the primary compression pulse by its surface reflected rarefaction, is obtained at the expense of generating bubble oscillations which give rise to secondary pressure pulses at every contraction of the gas bubble. These bubble pulses, whose form is shown later in Figs. 10 and 13, do not occur with shallow charges because the gas bubbles blow out into the atmosphere at their first expansion. They may be a useful contribution to total energy and penetration, but require special attention in order to improve resolution.

Despite the depth advantage, the dynamite source is still inefficient within the seismic band as shown in Fig. 8.

Reduction of the secondary pressure pulses radiated from a deep small charge of dynamite can be achieved in the field acquisition by surrounding the charge with a screen or cage, or by elongating the dynamite to become a thin line charge. Alternatively, correlation-type processing methods can aim to utilise the total energy radiated.

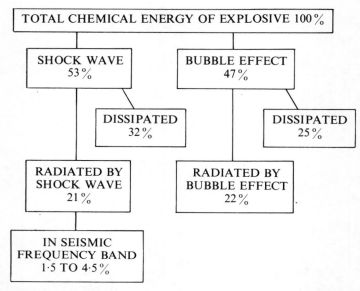

FIG. 8. Distribution of the total chemical energy in an underwater explosion of a high explosive. TNT (trinitrotoluene) density 1·5 g/cc. From Lavergne.[2]

THE LOW ENERGY ERA

In addition to the use of dynamite in these non-conventional forms, many other mechanical type non-dynamite sources, both explosive and implosive, were developed and operated in the early 1970s and referred to as low energy sources. They are described briefly in the following text.

Maxipulse® (Explosive Sound Source)

The Maxipulse system (Fig. 9) shoots 224 g charges of nitrocarbonitrate explosive 12 m below the surface at intervals of 12 s. Each 'Superseis'®† (explosive) charge is primed by the gun operator and propelled by fast flowing water along a hose, over the stern of the shooting vessel, to the firing depth. On reaching the gun at the end of the hose, the detonator is struck, the charge is ejected and explodes 1 s later several metres away from the gun. The source signature is recorded by a hydrophone near the end of the gun

® Trade mark and service mark of Western Geophysical Co.
®† Trade name of Hercules Inc.

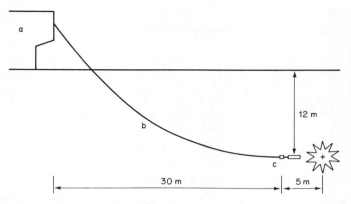

FIG. 9. Schematic of Maxipulse® system; a–shooting vessel, b–hose, c–gun transducer position. Gun transducer—−3 dB, 0·5 Hz to >10 kHz. ® Western Geophysical Co.

hose, and subsequently used to establish the time break and provide a record of the sequence of pulses produced by each shot.

The initial pressure front produced has an extremely fast rise-time. The gas bubble formed by this chemical explosion expands until the gas pressure falls to well below that of the surrounding water. It then contracts and expands in a series of bubble oscillations whose properties depend on charge size and water depth. The first bubble period is about 95 ms for a shot depth of 15 m and about 135 ms for a shot depth of 8 m. The charge weight is strictly controlled in manufacture so that changes in bubble period are mainly due to changes in shot depth.

A significant feature of the method is the design of a digital processing operator from the recorded near-field shot and bubble sequence signature. A typical signature is shown in Fig. 10 together with its amplitude spectrum. The operator aims to modify the recorded data to that which would have been produced by the simple single pulse shown in Fig. 11.

Flexotir® (Caged Explosive Sound Source)

The Flexotir system (Fig. 12) shoots a 50 g explosive charge simultaneously in each of two perforated spheres up to 15 m below the surface at intervals of 18 s. The explosive cartridge is jetted down a flexible hose by sea water at 100 psi pressure and detonated when it reaches the geometric centre of the

® Trade mark of Institut Français du Petrole.

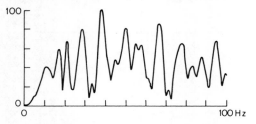

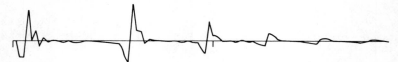

FIG. 10. Maxipulse® signature as recorded and its amplitude spectrum. ® Western Geophysical Co.

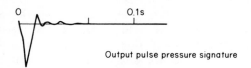

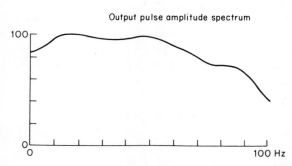

FIG. 11. Maxipulse® signature after data processing and its amplitude spectrum. ® Western Geophysical Co.

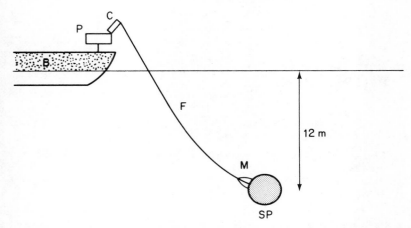

FIG. 12. Schematic of Flexotir® system; B–Recording boat, P–Pump, C–Loading head, F–Rubber hose, M–Firing device, SP–Sphere. ® Institut Français du Petrole.

2 in thick, 2 ft diameter cast iron sphere which is perforated with about 130 holes of 2 in diameter.

The initial pressure front is not appreciably affected by the cage but the gas bubble oscillations are severely damped, as the work performed in forcing water out and in through the holes in the sphere consumes kinetic energy. The bubble reduction effect is shown in Fig. 13 where the signatures of caged and uncaged 50 g charges at 13 m depth are compared in the 10–70 and 2–300 Hz bands.

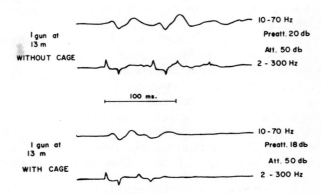

FIG. 13. Flexotir bubble attenuation by a 28% hole density cage. Recording distance 100 m below charge. From Lavergne.[2]

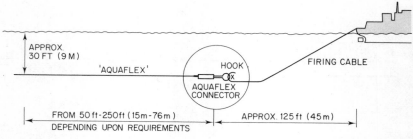

FIG. 14. The Aquaseis® system. ® Imperial Chemical Industries Ltd.

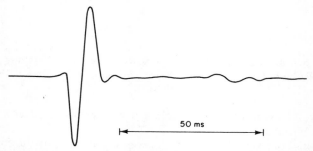

FIG. 15. Pressure signature of 30 m Aquaflex at 10 m depth, recorded in the 5–160 Hz band. From Jenyon.[3]

Aquaseis® (Linear Energy Source).

The Aquaseis system (Fig. 14) fires a 30 m length of Aquaflex®† seismic cord containing a core of petn (penta erythritol-tetranitrate) explosive of 1·5 lb weight at depths up to 12 m below the surface at intervals of 18 s. The cord is terminated with a connector which carries the detonator, and a ring clip which is attached to the firing line for loading the charge. When the connector reaches the firing line hook, the cord streams at the predetermined depth and is electrically detonated by means of a high voltage blaster via the sea water return path.

The progressive detonation of the line charge and the elongated nature of the bubble reduce the effective bubble amplitude and period, compared to the same weight of a point charge. The signature of a 30 m length of Aquaflex cord at 10 m depth in the 5–160 Hz band is shown in Fig. 15.

® Trade mark of Imperial Chemical Industries Ltd.
®† Trade name of explosive cord manufactured by Imperial Chemical Industries Ltd.

FIG. 16. Surface effect of 30 m Aquaflex fired at 25 ft depth.

Firing a 30 m length of Aquaflex at 8 m depth merely produces a fine spray above the surface as shown in Fig. 16.

Seisprobe®‡ (Gas Gun Sound Source)

The Seisprobe sleeve exploder system uses four units towed at 20–30 ft below the surface, each of which simultaneously fires by spark plug ignition (at 8–10 s intervals) a controlled mixture of propane and oxygen, which expands an elastic rubber sleeve surrounding the firing chamber. A single gun in its cradle on the ship's side is shown in Fig. 17. When the sleeve reaches maximum extension (Fig. 18(a)) the pressure inside it is substantially below the external hydrostatic pressure. This pressure differential, combined with the elastic restoring force of the sleeve, drives the sleeve inward until it contacts the inner cage (Fig. 18(b)). Combustion products are driven out and exhausted to the air through a valve set to

®‡ Service mark of Seismograph Service Ltd, Licensee of Esso Production Research Co.

FIG. 17. Single Seisprobe® gun. ® Seismograph Service Ltd, Licensee of Esso Production Research Co.

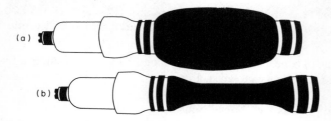

FIG. 18. Seisprobe® sleeve; (a) expanded and (b) contracted. ® Seismograph Service Ltd, Licensee of Esso Production Research Co.

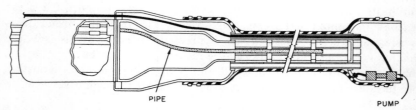

FIG. 19. Section of Seisprobe® gun. ® Seismograph Service Ltd, Licensee of Esso Production Research Co.

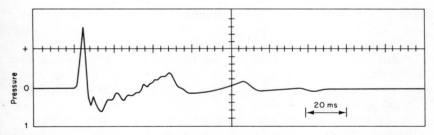

FIG. 20(a). Seisprobe® pressure pulse with 1·25 s gas fill. Source depth 20 ft.
® Seismograph Service Ltd, Licensee of Esso Production Research Co.

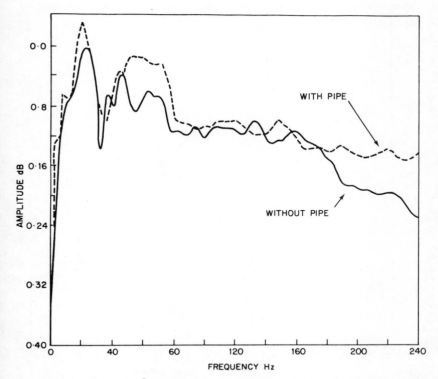

FIG. 20(b). Seisprobe® amplitude spectrum with and without pipe.
® Seismograph Service Ltd, Licensee of Esso Production Research Co.

open as the sleeve approaches its maximum extension. The fuel metering and detonation cycles are sequenced by a single control unit, and its gauges monitor combustion chamber performance and gas pressures.

When the system was first introduced, each gas gun required lifting clear of the sea water and purging of water condensate every 4 h. To overcome this inconvenience, jet pumps are now fitted to remove the water continuously.

A section of a single gas gun is given in Fig. 19. This shows a modification to improve the efficiency, by fitting a pipe to carry the burning front from the spark plug into the body of the gun. The near-field signature of a single gun at 20 ft with a 1·25 s gas fill is shown in Fig. 20(a). Its spectrum and the improvement due to the pipe are shown in Fig. 20(b). Peak pressure versus distance for a single gun is given in Fig. 21.

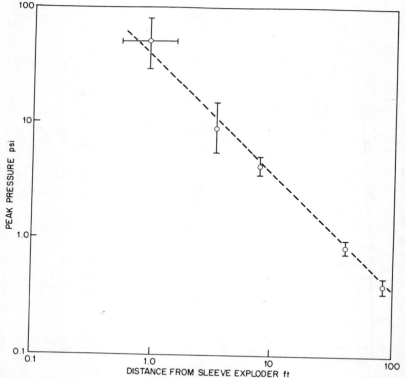

FIG. 21. Pressure versus distance for a single Seisprobe® at 20 ft depth. Fill time 1 s. ® Seismograph Service Ltd, Licensee of Esso Production Research Co.

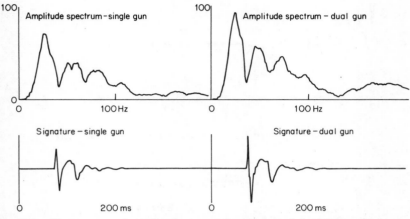

Fig. 22. Signatures and amplitude spectra for single and dual Seisprobe guns at 20 ft depth. Fill time 1·5 s.

Energy output is increased not only by increasing the gas fill time but by coupling single guns into dual or triple units alongside each other approximately 2 ft apart. Far-field signatures for single and dual guns are shown in Fig. 22 together with their amplitude spectra.

Propane and oxygen are normally carried aboard in gaseous or liquid form but oxygen can be provided from an oxygen concentrator plant, which sieves nitrogen from the air providing oxygen enriched air.

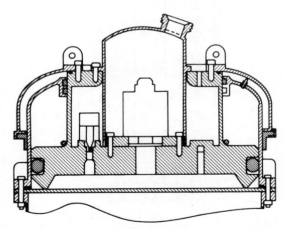

Fig. 23. Section of 24 in Dinoseis® unit. ® Sinclair Research Inc.

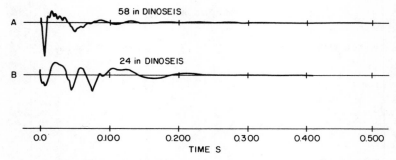

Fig. 24. Near-field signatures of 24 in and 58 in Dinoseis.

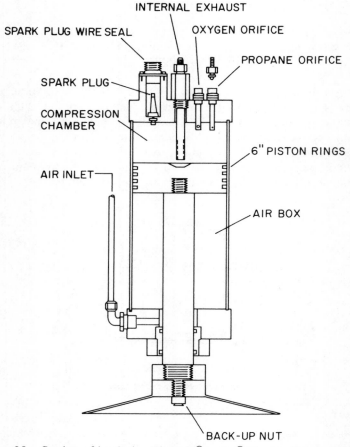

Fig. 25. Section of implosive Dinoseis® unit. ® Sinclair Research Inc.

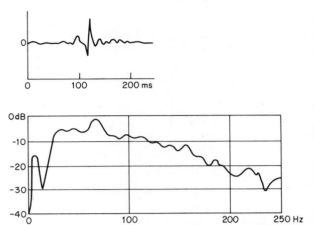

FIG. 26. Signature and amplitude spectrum for two Dinoseis® imploder units at 8 ft depth. 1·5 s fill. ® Sinclair Research Inc.

Dinoseis® (Gas Sound Source)

The Dinoseis gas exploder system uses one or more synchronised units, each of which consists of two interleaved case assemblies as shown in Fig. 23, which can slide on 'O' ring seals relative to one another. They are normally held together by a compressed air spring. The chamber immediately above the lower diaphragm is initially filled with an explosive mixture of propane and oxygen. When ignited by an electric spark, the increased gas pressure rapidly forces the two case assemblies into maximum extended positions. Thus the necessary high volume acceleration is imparted to the surrounding water. A 24 in diameter model weighs 1500 lb and a 58 in diameter model weighs 7000 lb. Their signatures are given in Fig. 24.

The Dinoseis gas imploder system uses one or more synchronised units, each of which fires by spark plug ignition a controlled mixture of propane and oxygen to drive a single plate through the water at high velocity, to form a cavitation bubble. The natural collapse results in a sharp acoustic pulse. A sectional view is shown in Fig. 25. The signature for two units at 8 ft depth recorded by a well geophone at 900 ft depth is shown in Fig. 26 together with its amplitude spectrum.

Hydrosein®† (Implosive Sound Source)

The Hydrosein system uses two units each of 9000 lb weight, fired 40 ft

® Trade mark of Sinclair Research Inc.
®† Trade mark of Western Geophysical Co.

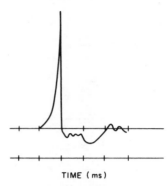

FIG. 27. Implosive portion of Hydrosein signature.

below the surface at 10 s intervals. High pressure air is moved into the piston chamber and the piston, piston rod and plate are accelerated downward away from the fixed upper plate (a distance of about 12 in) causing a cavity void between the two plates. Water rushing in to fill this void produces a strong implosive pulse as shown in Fig. 27.

Flexichoc® (Implosive Sound Source)
The Flexichoc system consists of two or more units (type F 123.20) towed at up to 15 m below the surface via an elastic nylon rope and control hoses allowing an 18 s firing interval. Each unit consists of a flexible envelope surrounding a pair of rigid circular plates set opposite each other and attached by jointed legs.

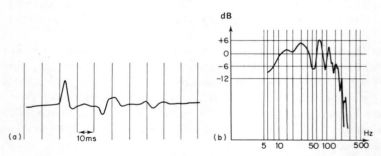

FIG. 28. Single Flexichoc® at 14 m depth; (a) signature, (b) amplitude spectrum.
® Institut Français du Petrole.

® Trade mark of Institut Français du Petrole.

The enclosure is enlarged to its maximum volume by a pump which injects compressed air at a slightly higher pressure than hydrostatic until the jointed legs lock the plates apart, after which the pump empties out the injected air and lowers the inside pressure to 100 mm of mercury. The locking mechanism is then released and the hydrostatic pressure on the walls of the housing abruptly reduces the volume to a minimum, generating an implosive pulse. The shock at the end of contraction is taken up by a spring.

The pressure signature from a single Flexichoc at 14 m depth in the record band 6–250 Hz is shown in Fig. 28, together with its amplitude spectrum. A later model (type FHC 50) in which oil pressure is used to force

FIG. 29. Single Flexichoc® unit. ® Institut Français du Petrole.

the plates apart is pictured in Fig. 29. Its signature at 5 m depth in a broader frequency record band is given in Fig. 30 together with its amplitude spectrum.

Vaporchoc® (Implosive Sound Source)

The Vaporchoc steam gun system (Fig. 31) consists of a single unit towed 7 m below the surface and fired every 8 s. Steam fed from the ship through an insulated pipe to the underwater storage tank is triggered into the water through a valve which opens for a set time, generally 40 ms. The steam bubble grows until injection is stopped. Then the steam begins to condense and the radius of the bubble begins to decrease due to hydrostatic pressure.

® Trade mark of Institut Français du Petrole.

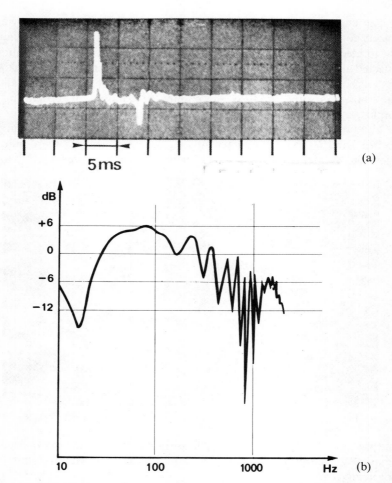

FIG. 30. Single Flexichoc® at 5 m depth; (a) signature, (b) amplitude spectrum.
® Institut Français du Petrole.

When the radius becomes very small, all the energy has been converted to kinetic energy in the inward flowing water, and due to spherical convergence a very high pressure is developed in the water near the inside 'wall' of the bubble. The high pressure implosion radiates acoustic energy and as there is no compressed gas to push the water outward, the bubble collapses completely and does not oscillate.

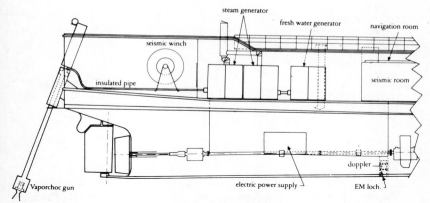

FIG. 31. The Vaporchoc® system. ® Institut Français du Petrole.

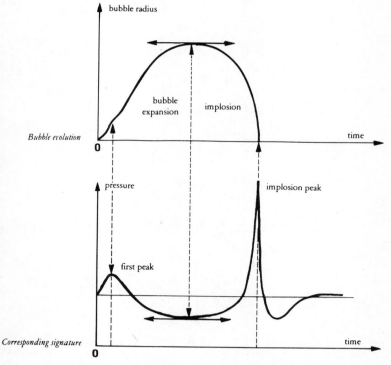

FIG. 32. Vaporchoc® signature and its relationship to bubble development. ® Institut Français du Petrole.

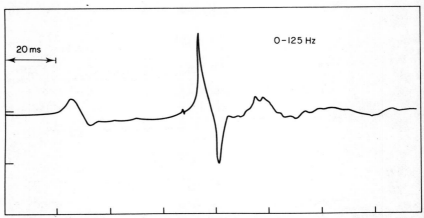

FIG. 33. Vaporchoc® far-field signature. ® Institut Français du Petrole.

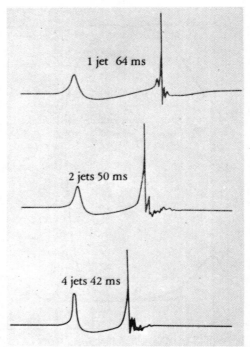

FIG. 34. Vaporchoc® near-field signatures for one, two and four jets. ® Institut Français du Petrole.

The initial expansion of the steam bubble, however, creates an expansion pulse prior to the implosion pulse and the signature is of the form shown in Fig. 32. A far-field signature is shown in Fig. 33. A hydrophone located near the gun records the signature for use in processing. Recent developments allow the monojet to be replaced by a multijet consisting of two, four or eight jets which give a choice of bubble period as indicated in Fig. 34.

PAR® (Air-Gun Sound Source)

The PAR air-gun is manufactured in several models covering a range of air chamber volumes up to 2000 in^3 and operating pressures up to 2000 psi. It provides an acoustic output by the explosive release of high pressure air directly into the surrounding water.

The method of operation is illustrated schematically in Fig. 35. On the right of the figure the gun is shown armed. Its upper control chamber and its lower discharge chamber are sealed by a triggering piston and a firing piston which, connected by a common shank, form a shuttle. A shipboard

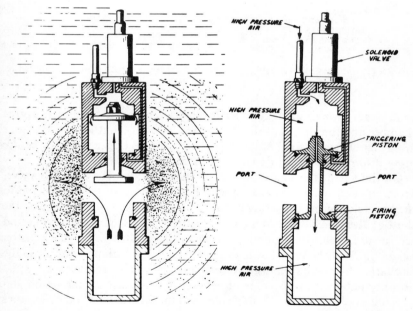

FIG. 35. Schematic PAR® air-gun operation. ® Bolt Associates Inc.

® Trade mark of Bolt Associates Inc.

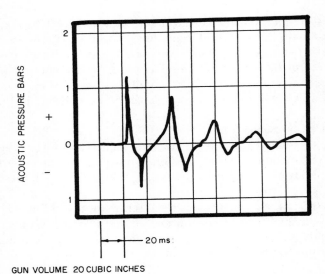

FIG. 36. Signature of 20 in³ PAR air-gun. Model 600B. From Bolt Associates Inc.

compressor supplies air at 2000 psi through an air hose to the upper control chamber and the air bleeds through an orifice in the shuttle into the lower firing chamber. The area of the triggering piston is larger than that of the firing piston and the net downward force seals the gun.

On the left of the figure the gun is shown firing. The solenoid valve, actuated electrically, allows a pulse of high pressure air to be delivered to the underside of the triggering piston. This upsets the force balance on the shuttle valve which opens at high velocity, reaches its maximum stroke and returns to its sealed position in a period of about 10 ms. The firing piston passes the four open ports at high velocity, and the high pressure air in the discharge chamber is explosively vented to the water producing an initial pressure pulse with a rise-time of between 1 and 5 ms, depending on gun size, followed by a series of bubble pulses of the form shown in the signature of Fig. 36.

Unlike the bubble pulses from chemical explosives whose periods decrease progressively, the periods of successive explosive air-gun bubble

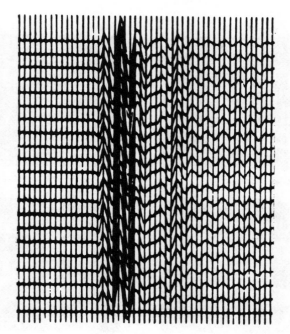

FIG. 37. Repeatability test on 10 in³ air-gun. Hydrophone 10 ft from gun. Timing lines 0·01 s. From Seismograph Service Corp.

pulses remain appreciably constant and may be estimated from the Rayleigh–Willis[4,5] formula

$$T = \frac{(PV)^{1/3}}{510\left(1 + \dfrac{H}{33}\right)^{5/6}}$$

where T is the period in s, H is the gun depth in ft, V is the volume of the gun in in³ and P is the firing pressure in psi. The repeatability of a 10 in³ air-gun signature is indicated by the duplication from twenty shots shown in Fig. 37.

Reduction of the secondary bubble pulses from individual air-guns is made possible by delaying the release of part of the air in the firing chamber, so that by feeding its own bubble its rate of bubble collapse is slowed. The firing chamber is divided into two parts by means of a plate with a central orifice. By adjusting the volume division and the size of the orifice, it is possible to vary the amount of the bubble suppression with some loss of initial pulse peak pressure, as shown in Fig. 38.

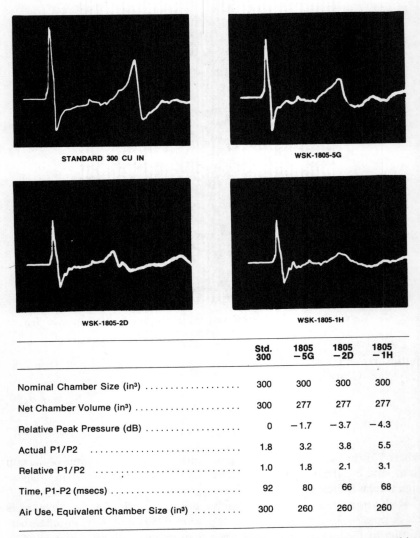

FIG. 38. Outputs of PAR air-gun model 1500C with several waveshape kits. Oscillograms not to same scale. Refer to table for comparison. P_1 and P_2 refer to the first and second pressure peaks. From Bolt Associates Inc.

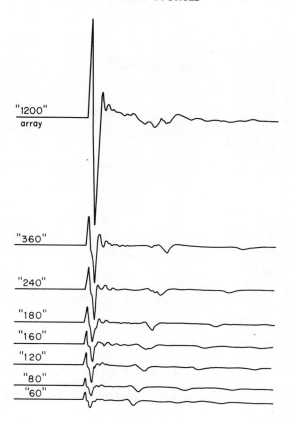

FIG. 39. Signature of an air-gun array and the signatures of the 7 individual non-interacting groups of which the array is composed.

Reduction of the secondary bubble pulses from a number of guns is achieved by simultaneously firing guns at the same depth with different chamber volumes, so that their individual initial pulses add in phase but their bubbles do not. This results in an improvement in initial pulse–bubble pulse ratio in the vertical direction, as shown in Fig. 39. Here, the individual far-field signatures of seven different volumes at 23 ft depth are summed to produce this 1200 in^3, far-field, synthetic, non-interacting array signature.

Using selected combinations of volume, pressure, depth, spatial separation and firing time from arrays of guns, it is possible to radiate almost any type of signature in any preferred direction.

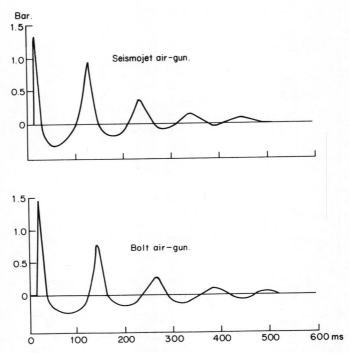

FIG. 40. Comparison of the Seismojet® 125 in³, 8000 psi air-gun with the Bolt 300 in³, 2000 psi air-gun. Air-gun and detector distance 4 ft, depth 15 ft. ® Trojan US Powder. Manufactured and distributed by Dresser SIE Systems.

Seismojet® (High Pressure Air-Gun Sound Source)

The Seismojet system employs a number of explosive air-guns with chamber volumes of 40–125 in³ operated at a pressure between 6000 and 8000 psi. The system employs an electro-mechanical sequencing technique to attenuate the bubble by a controlled air discharge after the initial wave front has been generated, which in effect maintains the bubble and prevents its implosion. Near-field comparison signatures between a 125 in³ gun at 8000 psi and a 300 in³ gun at 2000 psi for a depth of 15 ft are shown in Fig. 40. Bubble attenuation effect is illustrated in Fig. 41. The air injection ports are self-cleaning allowing operation in shallow water or marsh areas.

® Trade name of Trojan US Powder. Manufactured and distributed by Dresser SIE Systems.

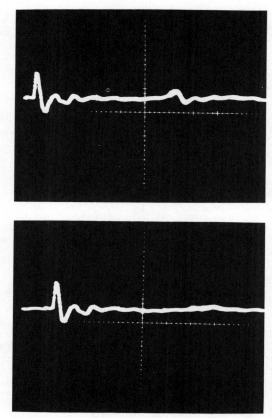

FIG. 41. Seismojet® bubble attenuation. Top: bubble effect shown; bottom: bubble effect eliminated. ® Trojan US Powder. Manufactured and distributed by Dresser SIE Systems.

Simplon®† (Implosive Sound Source)

The Simplon water gun is manufactured in several models covering a range of air chamber volumes up to 915 in^3 and operating pressures up to 3500 psi. It is basically an axial flow air-gun coupled to a piston and cylinder assembly. It produces an acoustic pulse by implosion of a void created under water. The method of operation is illustrated in Fig. 42.

The air-gun is brought to pressure. The piston is maintained against the

®† Trade mark of Société Pour le Développement de la Recherche Appliquée (SODERA).

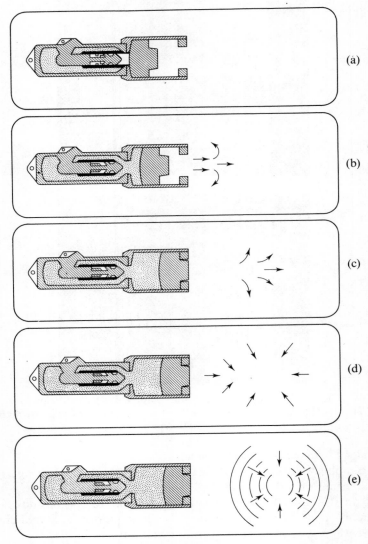

FIG. 42. Schematic operation of water gun. From SODERA.

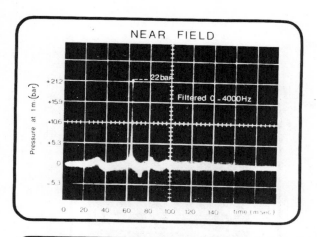
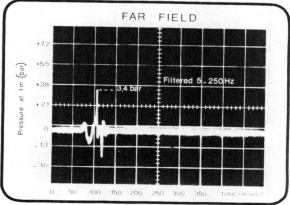
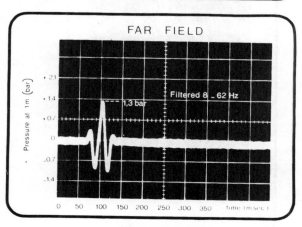

FIG. 43. Mica® water gun 120 in³ near-field and far-field signatures. ® SODERA.

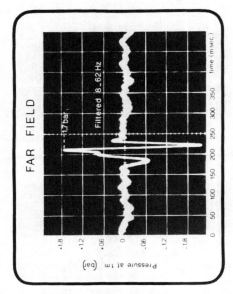
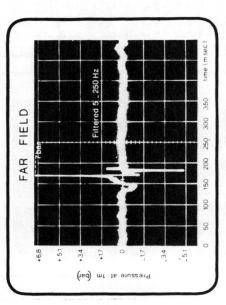

FIG. 44. Cassios® water gun 540 in³ far-field signatures. ® SODERA.

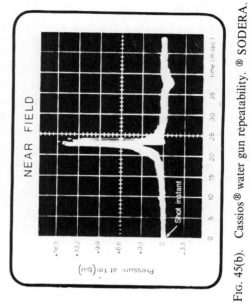

Fig. 45(a). Mica® water gun repeatability. ®SODERA.

Fig. 45(b). Cassios® water gun repeatability. ® SODERA.

air-gun outlet port by hydrostatic pressure and the cylinder is filled with water in front of the piston (Fig. 42(a)). The air-gun is triggered by an electrical solenoid and its high-pressure air propels the piston at high velocity (100–200 m/s), expelling the water contained in the cylinder in the form of a water plug (Figs. 42(b) and (c)). As the piston is rapidly decelerated, the moving water mass is separated from the piston and a cavity is formed by inertia (Fig. 42(d)). This cavity implodes (Fig. 42(e)) creating the acoustic signal whose energy is proportional to the kinetic energy of the water plug. At the end of the stroke the piston stops, the air-gun port closes and the air behind the piston in the cylinder is vented to the surface. As the air-gun refills with air, the piston is pushed back against the air-gun port by hydrostatic pressure.

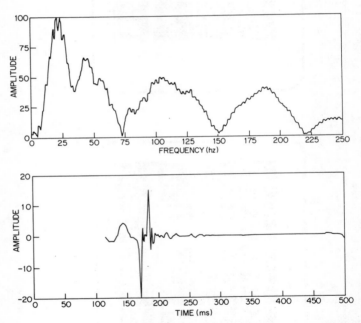

FIG. 46. Cassios water gun 540 in³ far-field signature and amplitude spectrum.

The wide-band near-field signature and filtered far-field signatures of a 120 in³ Mica® water gun fired at about 10 m depth with 2100 psi air pressure are shown in Fig. 43. Far-field filtered signatures of a 540 in³

® Trade name of SODERA.

Cassios®‡ water gun fired at the same depth and with the same pressure are shown in Fig. 44.

The repeatability of the 120 in^3 Mica water gun over ten successive shots in the 0–1 kHz band is shown to be ± 1 ms in Fig. 45(a) compared to $\pm 1\cdot 5$ ms for the 540 in^3 Cassios over twenty successive shots in Fig. 45(b). The far-field signature from a single 540 in^3 Cassios water gun fired at 10 m depth with 2000 psi air pressure is shown in Fig. 46 together with its amplitude spectrum. Without the Simplon attachment the gun may be used as an explosive air-gun.

Vibroseis® (Continuous Sound Source)

The Vibroseis system makes use of four hydraulic vibrators towed 35–40 ft below the surface. The water piston of each vibrator is driven by a hydraulic ram whose hydraulic pump, controlled by a high rate servo valve, follows the selected swept frequency signal over a range of frequencies from 10–100 Hz for a sweep time of about 5 s. The relatively long radiated 'chirp' signature is correlated with the recorded data to produce the seismic record.

The vibrator shown in Fig. 47 has a 4 ft diameter water piston and weighs 5000 lb. Its response is shown in Fig. 48. A sweep of 10–40 Hz over 5 s with a 5 s listening time and three composites per shot-point is typical and is illustrated in Fig. 49.

Magnetic Air-Gun®† (Low Pressure Air-Gun Sound Source)

The Bolt magnetic air-gun system consists of two or four guns, each of weight 650 lb, and of the type shown in Fig. 50, towed individually astern at 30–40 ft depth and fired at 10 s intervals by electric solenoid operation. Firing pressure is 500 psi with individual chamber volume, 1550 in^3. Single gun signatures are given in Fig. 51 for both normal (resonant) mode and wave-shaped (wide-band) mode.

Electrical Discharge Sources—Sparkers

The spark discharge through one or more electrodes of a battery of condensers precharged for a few seconds at several kilovolts provides a marine seismic source for shallow-penetration high-resolution surveys. When a high voltage is discharged from an electrode under water, a bubble is formed. The rapid expansion of this bubble or plasma provides the initial acoustic energy pulse. Its rapid collapse gives rise to a secondary pulse

---

®‡ Trade name of SODERA.
® Trade mark and service mark of Continental Oil Co.
®† Trade name of Bolt Associates Inc., Licensee of Mobil Oil Corp.

FIG. 47. Vibroseis high power transducer. From Seismograph Service Corp.

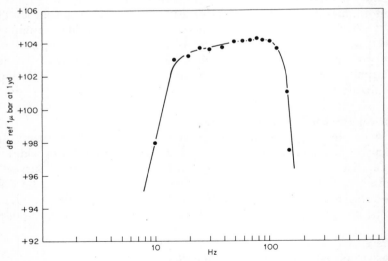

FIG. 48. Vibroseis. Single vibrator response. From Seismograph Service Corp.

MARINE SEISMIC SOURCES

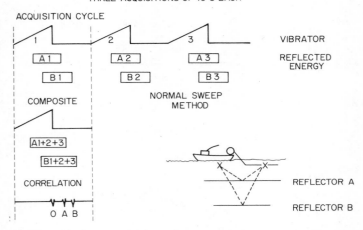

FIG. 49. Vibroseis data acquisition. Normal sweep method. From Seismograph Service Corp.

FIG. 50. Magnetic Air-gun.® ® Bolt Associates Inc., Licensee of Mobil Oil Corp.

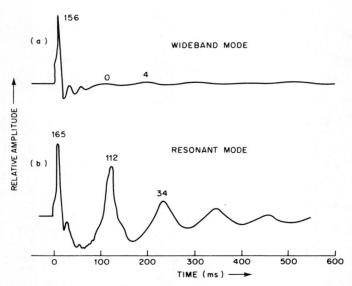

FIG. 51. Magnetic Air-gun® signatures; (a) wide-band mode, (b) resonant mode. ® Bolt Associates inc., Licensee of Mobil Oil Corp.

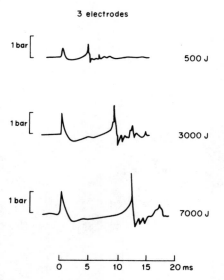

FIG. 52. Sparker signatures from 3 E.G. & G. Environmental Services electrodes at 5 m depth. Recorded 0–15 kHz. From Cassand and Lavergne.[6]

whose peak pressure is often greater than the initial pulse, as shown in Fig. 52.[6]

There is only one bubble pulse, since steam condenses to water during the bubble expansion and contraction cycle and there is little gas remaining. The collapse terminates in a sharp water slap. By increasing the number of electrodes whilst controlling electrode separation and energy per electrode, shorter high ratio signatures may be achieved, as shown in Fig. 53.[6]

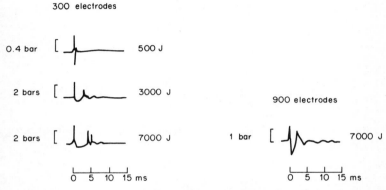

FIG. 53. Sparker signatures from 300 and 900 Institut Français du Petrole electrodes at 5 m depth. Recorded 0–15 kHz. From Cassand and Lavergne.[6]

Minisleeve ® (Gas Sound Source)

A recent development of the sleeve exploder intended for shallow-penetration high resolution work is the Minisleeve. This miniaturised version generates a short simple pulse (Fig. 54) with a broad spectrum bandwidth (Fig. 55) from twelve sleeves each 5 cm diameter and 30 cm long, which are mounted on a sled 3 m × 1·5 m towed between pontoons at a depth of 0·5 m.

The sleeves are supplied with propane and oxygen gases and are purged and controlled in pairs through a single tow member 45 m long. They can be fired sequentially at a minimum time interval of 300 ms to give dense sub-bottom coverage, or simultaneously every 1·8 s to provide maximum penetration.

® Trade name of Esso Production Research Co.

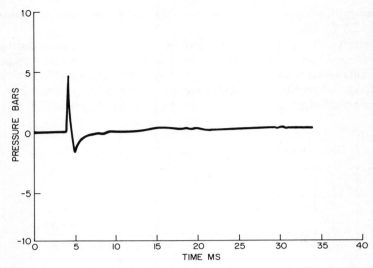

FIG. 54. Minisleeve® signature. ® Esso Production Research Co.

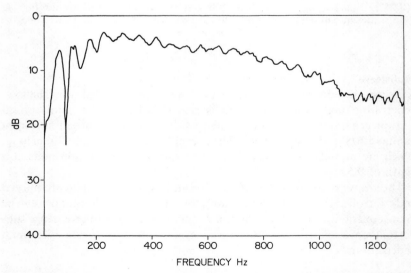

FIG. 55. Minisleeve® amplitude spectrum. ®Esso Production Research Co.

THE TECHNICAL ASSESSMENT OF SOURCES

The intrinsic energy levels for the various types of sources discussed are compared in Table 1, column (a). Their measured bubble periods when operated as single unit sources at 30 ft depth are given in column (b). Values from (a) and (b) are plotted as period energy points in Fig. 56 and compared with the Rayleigh–Willis[4,5] curve in the manner outlined in *Seismic energy sources handbook*.[7] The closer the plotted points approach the Rayleigh–Willis curve the larger is the proportion of intrinsic energy available for bubble formation and acoustic radiation. The pressure–time curve or signature is more instructive and when accurately recorded by a calibrated detector within a specified frequency band, allows absolute source comparisons to be made on a peak pressure basis, as well as providing information on period and form.

The maximum positive peak pressure is measured for near-field or mid-field signatures as in Fig. 36, whereas maximum positive to maximum negative peak to peak pressure swing is measured for far-field signatures which include the water surface reflection as in Fig. 46.

The ratio (P_1-P_2) relating primary peak or peak to peak pressure to maximum secondary pressure peak or peak to peaks in the signature provides a useful form factor especially for arrays in the far-field.

Positive primary peak pressures measured in bars within the frequency band 0–62 Hz and referred back to a distance of 1 m from the source are given in column (c) of Table I.

Single unit sources avoid synchronising problems, but except for dynamite are weak and, except for Aquaseis, are omnidirectional.

Non-dynamite low energy sources deploy multiple units to increase total source strength but other advantages accrue. Smaller sources can be fired more often to increase the fold of stack. Energy division itself increases radiated efficiency although low frequency response suffers. The arraying of a number of units allows directional properties to be introduced and the possibility of cavitation at the surface is reduced. If the far-field signature of the multiple units is to remain identical in shape and increase only in strength to provide maximum possible energy, the individual units must duplicate. They must be at the same depth and accurately synchronised, and they must be spaced far enough apart so that interaction does not prevent linear addition.

For impulsive sources or for the impulsive events in their signatures, interaction is related to peak pressure or bubble radius and relatively short separation distances avoid impulsive interaction. However, for continuous,

TABLE 1
COMPARISON OF INTRINSIC ENERGY, PERIOD AND PEAK PRESSURE FOR SINGLE UNIT SOURCES

	(a) Intrinsic energy K ft lbs	(b) Period at 30 ft ms	(c) Primary pulse 0–62 Hz Barmetre peak
50 lb Dynamite	61 000	500	
10·5 lb Dynamite	12 000	300	13
1 lb Dynamite	1 200	105	
0·25 lb Dynamite	304	82	
1·5 lb Aquaseis	2 000	54	
224 g Maxipulse	590	95	4
50 g Dynamite	145	69	2·5
50 g Flexotir	145	48	2·1
1·5 s Seisprobe	90	40	0·4
Bolt 1 in^3 Airgun 2 000 psi	1	14	0·2
Bolt 10 in^3 Airgun	10	30	0·6
Bolt 20 in^3 Airgun	20	39	0·7
Bolt 120 in^3 Airgun	120	70	1·2
Bolt 300 in^3 Airgun	300	100	1·7
Bolt 2 000 in^3 Airgun	2 000	165	4
125 in^3 Seismojet 8 000 psi	480	94	1·5
125 in^3 Seismojet W/S 8 000 psi	480	70	1·2
Sodera 120 in^3 Watergun 2 100 psi	126	26	1·3
Sodera 550 in^3 Watergun 2 100 psi	578	45	1·7
Sodera 915 in^3 Watergun 2 000 psi	920	55	
Sodera 915 in^3 Airgun 2 100 psi	920	120	2·9
Sodera 125 in^3 Airgun 3 500 psi	210	80	
Vaporchoc	943	64	2·5
Flexichoc	36	31	0·6
1·7 s Implosive Dinoseis	40	30	0·9
Hydrosein	64	38	
0·5 s Minisleeve	0·7	8	
3 kJ Sparker (3 electrodes)	2·2	5·5	

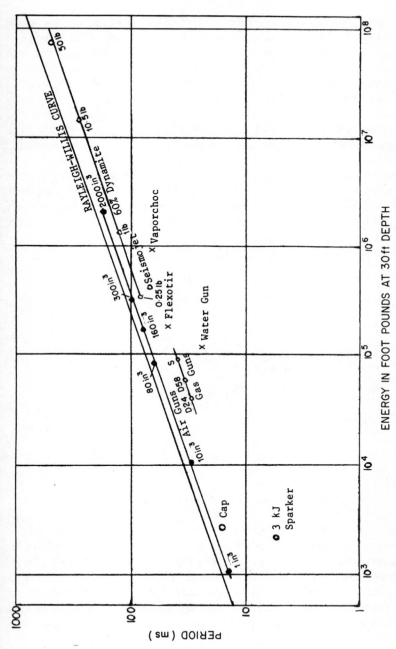

FIG. 56. Period energy points for single unit sources at 30 ft depth.

long signature type sources or for the bubble events of explosive sources, interaction is related to radiated wavelength and much larger separation distances are required to avoid wave interaction. Impulsive interaction is partially self-destructive and the total radiation of energy is reduced by turbulence, whereas wave interaction is an interference phenomenon and only the timing of radiated energy is affected. Energy division and interaction may thus be used for radiated signal shaping, but the interrelated effects of interaction and synchronisation together with depth and ghost effects require measurement in the far-field to establish signature and repeatability.

The effect of the free surface of the water on a near vertical radiated pulse of compression is to reflect it without loss of amplitude or change of form, but with a change of sign as a pulse of rarefaction giving the water a pressure less than hydrostatic, as shown in Fig. 57. If the pulse at the surface is large

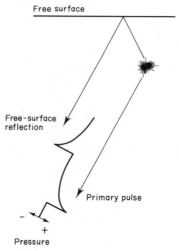

FIG. 57. Illustration of the free surface effect.

enough, the resultant rarefaction will exceed the tension strength of the water causing cavitation and limiting the negative pulse amplitude. The down-going doublet then appears as Fig. 58(b) instead of as Fig. 58(a). The time between primary pulse and free surface reflection decreases as source depth decreases, until the free surface reflection begins to erode the primary pulse and finally cancels it completely at the surface. The free surface reflection effect is more severe on secondary pulses, because they are

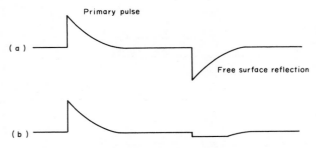

FIG. 58. Illustration of the cavitation effect.

generally broader pulses of lower frequency than the primary pulse for a given depth.

As depth and hydrostatic pressure increase, bubble periods decrease. The wide-band, 0–2·5 kHz, analogue far-field signatures for a single 120 in^3 air-gun fired at 2000 psi are displayed in Fig. 59. They illustrate the changes in bubble period and free surface reflection time effects which occur as gun depth increases from near surface to 120 ft. The detector depth remains constant at 320 ft. Their digitally recorded, 0–250 Hz, far-field signatures and amplitude spectra are displayed in Fig. 60 together with information relating primary peak amplitudes and total radiated energy. The ratio P_1–P_2 refers to primary to secondary peak to peak ratio, whilst the energy units given are proportional to the areas under the pressure signature curves.

The effect of synchronising error on the far-field sum of two non-interacting, 160 in^3 air-guns is illustrated in Fig. 61. The signatures and spectra are synthesised from a single 160 in^3 signature, recorded 0–250 Hz, and added to itself with the specified delays.

The decreasing effect of interaction with increasing separation distance for two 120 in^3 air-guns is shown by the far-field 0–250 Hz, signatures, spectra and related information in Fig. 62. The effect of synchronising error on the far-field sum of two interacting 120 in^3 air-guns is illustrated by the wide-band 0–2·5 kHz signatures in Fig. 63 and their 0–250 Hz signatures, spectra and related information in Fig. 64.

A GENERAL APPRAISAL OF SOURCES

Dynamite at shallow depth is very wasteful of non-radiated energy and a large proportion of the remaining radiated energy is produced at high

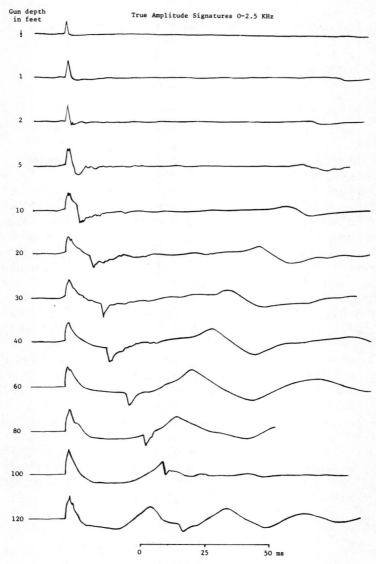

FIG. 59. Far-field 120 in³ air-gun signatures, 0–2·5 kHz.

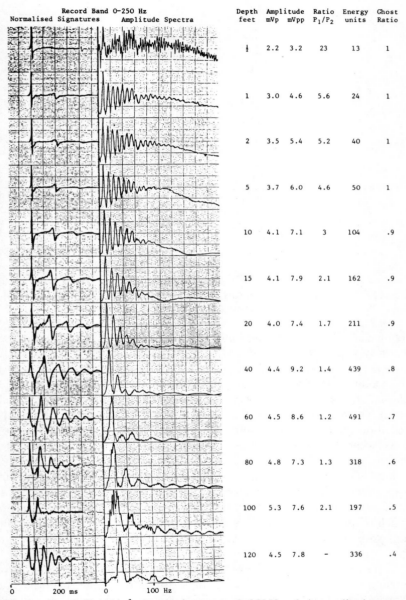

FIG. 60. Far-field 120 in³ air-gun signatures, 0–250 Hz; their amplitude spectra and related data. 1 energy unit ≃ 100 ft lb.

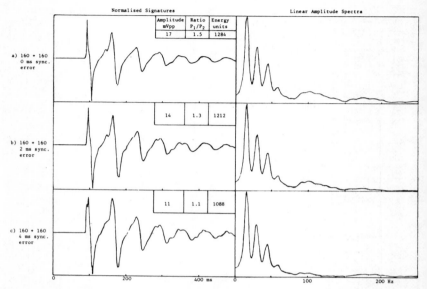

FIG. 61. The effect of synchronising error on two non-interacting 160 in³ air-guns. 1 energy unit ≃ 100 ft lb.

frequencies above the seismic band, resulting in high fish kill. Dynamite at depth is much more efficient but still produces a large proportion of its radiated energy at frequencies above the seismic band.

Dynamite is expensive, has a limited repetition rate and other disadvantages related to logistics and safety. Politically, it is 'dynamite'! However, Maxipulse did provide a low energy dynamite source during the development period of the mechanical type sources. Only in transition zones where low energy sources are unmanageable or high repetition rates impractical, is dynamite in use where permits allow.

Implosive sources, such as Hydrosein, Dinoseis, water gun and Vaporchoc also radiate a high proportion of energy at high frequencies above the seismic band, and are deficient in low frequency energy, most of which in any case is associated with the water pulse and not the primary implosive pulse. A large proportion of intrinsic energy is used to accelerate the water and produce this water pulse.

Vaporchoc is fixed to the vessel stern and has no depth compensation for pitching and rolling. It operates as a single source; it is limited in output and is omnidirectional. The other implosive type sources which deploy multiple units are subject to synchronising errors arising from the relatively long

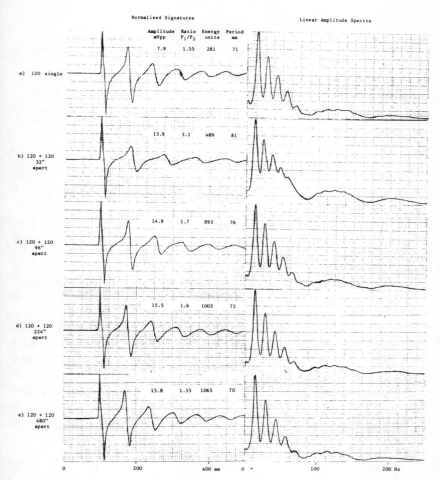

FIG. 62. The decreasing effect of interaction with increasing separation of two 120 in³ air-guns. 1 energy unit ≃ 100 ft lb.

time period which occurs between initiation and the primary implosion pulse.

Gas exploders are relatively weak individually and whilst accurate synchronising of multiple units is readily achieved, duplication of individual unit signatures, especially of the secondary pulse, is subject to sleeve elasticity or ageing of the sleeve, accurate control of gas mixtures and efficient purging. The period between primary and secondary pulses

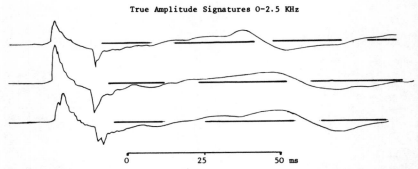

FIG. 63. The effect of synchronising error on two interacting $120\,\text{in}^3$ air-guns, 0–2.5 kHz.

remains reasonably constant over the operating ranges for gas fill and depth. The secondary pulse amplitude increases with increase in depth, reducing primary to secondary ratio. The supply of large quantities of propane gas is problematical in some parts of the world. The most recent Minisleeve development will surely be preferred to the sparker for future sub-bottom profiling surveys.

The air-gun is the most efficient of all the sources since most of its radiated energy is contained within the seismic band. Despite the fact that more than half of this energy is bubble energy, it is the most flexible and versatile, and is rapidly becoming the most favoured.

Arrays of air-guns may be designed to radiate in the near vertical

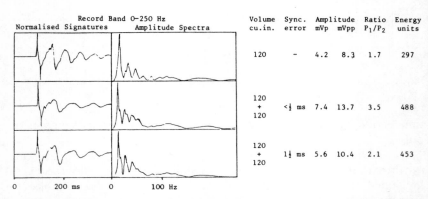

FIG. 64. The effect of synchronising error on two interacting $120\,\text{in}^3$ air-guns, 0–250 Hz; their amplitude spectra and related data. 1 energy unit $\simeq 100\,\text{ft lb}$.

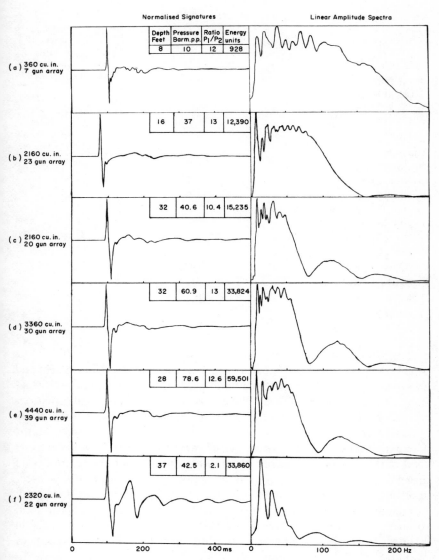

FIG. 65. Air-gun array signatures, 0–250 Hz; their amplitude spectra and related data. 1 energy unit ≃ 100 ft lb.

TABLE 2
COMPARISON OF SOURCE SYSTEMS AT 10 M DEPTH, 0–250 Hz

System	Radiated pressure Barm. p.p.	Primary to secondary ratio
1 gas gun	1·4	2
4 gas guns (4 × 1)	5	2
8 gas guns (4 × 2)	9	3
12 gas guns (4 × 3)	12	4
Maxipulse	10	0·9
Aquaseis		3
Flexotir 1 gun	5	4
Flexotir 2 guns	9·4	4
Airgun 1 gun 120 in^3	3·5	1·4
Airgun 7 guns 360 in^3	10	6
Airgun 20 guns 2 160 in^3	40	10
Airgun 39 guns 4 440 in^3	80	13
Watergun 11 guns 1 150 in^3	17	3
Flexichoc 1 gun	1·0	
Flexichoc 10 guns	8	
Vaporchoc 1 gun	6	3
Dinoseis Imp. 1 gun 1·7 s	2·25	
1 gun 3·5 s	4·5	3
2 guns 3·5 s	8·5	

direction, giving far-field signatures such as those recorded 0–250 Hz and shown in Fig. 65 together with their amplitude spectra and related information. Although other considerations such as near sea-bed and local conditions require attention, generally the high ratio signature of Fig. 65(a) with its higher frequency spectrum would be used for very shallow target depths where high repetition rates are required, whilst Fig. 65(b) would cater for shallow target depths with normal repetition rates. Generally, for moderate target depths the high ratio signatures of Fig. 65(c) and (d) would be used, and for deep target depths the signature of Fig. 65(e) would be preferred. For maximum energy within a given volume, bubble energy may be utilised to give a low ratio signature with a low frequency narrow band peaked response as in Fig. 65(f), in order to obtain penetration at the expense of resolution.

Some currently operating source systems with their multiple unit arrays have their peak to peak primary radiated pressures and primary to secondary pulse ratios compared in Table 2.

THE STATE OF THE ART IN CONTROL AND MEASUREMENT

The developments which have taken place since the introduction of the various types of sources have been related to improving the efficiency of the basic unit, increasing the number of units deployed and improving the control of the increased number deployed, to produce a larger amplitude, more repeatable and better defined total source signature.

Improved repeatability of the radiated far-field signature results from better depth control and improved synchronising. In practice, in moderate sea states, source units with near vertical tow arrangements have poor depth control. Near horizontal tow resulting in better decoupling of the source from vessel movements is preferred, with flexible links between individual units and streamlined floats, to reference the in-line array to the surface. Depth is monitored continuously at a number of points along the arrays. Good depth control reduces variations in fundamental frequency as well as travel-time differences of both primary and free surface reflections in the preferred, near vertical, direction.

Absolute synchronisation of a number of individual source signatures requires perfect duplication from initiation for each source. Effective synchronisation is the controlled timing of the peaks of the primary events within the limits of the individual source's own repeatability. Control circuitry is available to allow initiation in steps of 0·25 ms.

In practice, the ability of individual sources to remain within a chosen time window for repeatability, between maintenance periods, is greatly improved by initiation control and monitoring. This control is effected by a monitor on each gun, which detects an event occurring after initiation, and more closely related to the peak of the primary event, to adjust the initiation time to improve synchronisation automatically on a statistical basis within a given time window, and if this cannot be achieved, switches off the offending gun or allows it to be replaced. For non-interacting units, hydrophones are used as monitors of individual near-field signatures, as well as for synchronising control.

For interacting sources, especially air-guns, recent developments offer a number of devices unaffected by interaction, which detect a specific event between initiation and radiated peak pressure for use in control. The simplest of these and probably the most effective is a probe alongside the air-gun port which actually detects the emerging bubble of air. Its accuracy is indicated on the top trace of Fig. 66 which shows the superpositioning of fifteen probe detector pulses from the consecutive firings of an 80 in^3 air-gun.

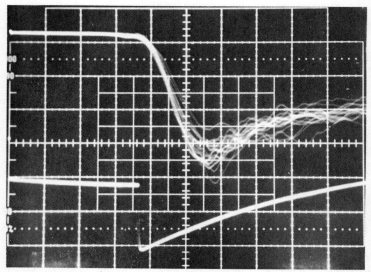

FIG. 66. Top trace; 15 probe detector pulses. Horizontal scale 0·2 ms per division.

Until recently, measurements of far-field source signatures were carried out off-line and not under actual operating conditions but at slow speed or stationary, mainly due to the difficulties of towing and positioning a reasonably quiet detector hydrophone far below the centre of the source array, at water speeds of up to 6 knots. Such source signatures were usually produced as standard signatures for a number of source depths and represented the desired on-line source signature for the chosen depth.

The growing need for analysis and verification of on-line source signature control, together with the improvement in data quality resulting from the use in processing of such standard signatures, motivated the development of the Sentinel® far-field recording system shown in Fig. 67. This system measures the depth of its far-field hydrophones and records their signatures, whilst towing at up to 150 m below the sea surface and at up to 6 knots water speed with a low tow noise characteristic.

Two hydrophones are normally towed. One for calibration and on-line quality control whose signature is displayed wide-band, in analogue form, on a fibre-optics oscilloscope. This may be recorded directly onto ultra-violet paper as in Fig. 67(a) which shows the unfiltered signature from a 3360 in^3 array. The second hydrophone, whose response duplicates the

® Service name of Seismograph Service Ltd.

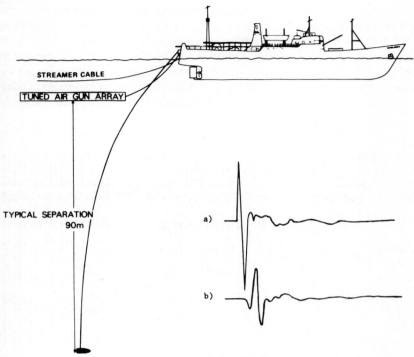

FIG. 67. Sentinel far-field recording system.

response of the hydrophones in the streamer cable, outputs its on-line far-field signature to take the same route through the seismic recording amplifiers as the data, to be monitored as in Fig. 67(b) and to be recorded on tape for every shot.

Prior to recording seismic data, the signatures may be analysed to optimise the primary to bubble ratio and the spectral content of the transmitted pulse. Improved array signatures may be developed. During the recording of seismic data, the signatures may be used to ensure correct operation of the air-gun array and provide quality control. During processing, the signatures may be used for deterministic deconvolution, to convert them to minimum phase to enhance Wiener deconvolution. Alternatively, pulse compression may be applied within the signature deconvolution where conventional spectrum whitening fails. The effects of shot to shot variations in the source signatures mainly due to sea state may be removed from the data.

The restrictions on the use of the far-field recorded signature imposed by water depth to sea-bed and required source depth, are set by the minimum requirement to record one full period of the source signature without interference from the bottom reflection, at a range from the source which prevents travel path differences for both primary and free surface reflections, due to geometry from introducing apparent synchronising errors. In water depths of 175 m with the source at 10 m and the detector

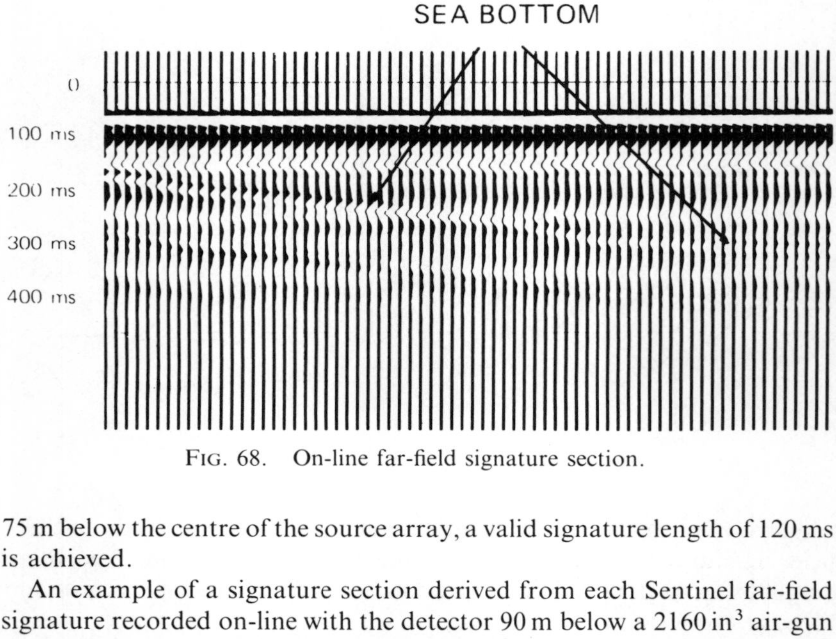

FIG. 68. On-line far-field signature section.

75 m below the centre of the source array, a valid signature length of 120 ms is achieved.

An example of a signature section derived from each Sentinel far-field signature recorded on-line with the detector 90 m below a 2160 in^3 air-gun array is shown in Fig. 68. The signature is repeatable. The sea-bed reflection can be seen progressively shortening the usable signature length.

Control and measurement of the radiated far-field signature from in-line arrays of low energy sources such as air-guns are the current state of the art. High ratio, high energy, repeatable signatures which approach the theoretical spike are being radiated, controlled and measured in the far-field on-line, in deep water, and are capable of being reasonably predicted in shallow water. Areal arrays designed to increase radiated efficiency and improve directivity will emerge, and as synchronising of implosive sources improves, a rising frequency response characteristic may well be incorporated to compensate for natural earth losses.

ACKNOWLEDGEMENTS

The author wishes to thank the Directors of Seismograph Service (England) Ltd for permission to publish this contribution, without associating them with any of the opinions or conclusions that are reached.

REFERENCES

1. JAKOSKY, J. J. and JAKOSKY, J., JNR, Characteristics of explosives for marine seismic exploration, *Geophysics*, **21**, No. 4, 1956.
2. LAVERGNE, M., Emission by underwater explosives, *Geophysics*, **35**, No. 3, 1970.
3. JENYON, M. K., *Experimental tests and field results obtained with underwater line charges*, paper presented at the 30th meeting of the European Association of Exploration Geophysicists (EAEG), Salzburg, Austria (unpublished).
4. RAYLEIGH, LORD J. W., On the pressure developed in a liquid during the collapse of a spherical cavity, *Philosophical Magazine*, **34**, 1917.
5. WILLIS, H. F., *Underwater explosions, time interval between successive explosions*, British Report WA 47.21, 1941.
6. CASSAND, J. and LAVERGNE, M., High resolution multi-electrode sparker, *Geophys. Prospecting*, **18**, 1970.
7. KRAMER, F. S., PETERSEN, R. A. and WALTER, W. C., *Seismic Energy Sources Handbook* 1968, p. 40, Bendix United Geophysical Corp., Pasadena, Ca, USA, 1968.

Chapter 6

GRAVITY AND MAGNETIC SURVEYS AT SEA

L. L. NETTLETON

Rice University, Houston, Texas, USA

SUMMARY

This chapter gives a short description of the history, instruments, corrections, navigation and data processing used in modern gravity and magnetic measurements at sea. A short section on interpretation includes maps and profiles from a marine survey.

The theory, instrumentation and recording of the La Coste and Romberg stabilised platform gravity meter are given in considerable detail. Brief mention is made of several other instruments which have been developed.

Special attention has been given to the various corrections to the data and the manner in which they are recorded on magnetic tape. Navigation systems and digital recording of their outputs are covered.

It is shown that the entire development of gravity measurements at sea depends on, first, the digital recording of essential data and, second, on the rather extensive computer processing of these records that takes place both on the ship and in the office. The section on interpretation covers some aspects which are different from those on land. It is illustrated by gravity and magnetic maps and profiles from an actual survey at sea.

INTRODUCTION

In the last twenty years or so the measurement of gravity and magnetics at sea has developed into a practical field operation able to measure gravity to about one milligal and magnetics to one or two gamma. This has resulted from two independent developments; the stable platform gravity meter and the proton precession magnetometer.

The stable platform gravity meter resulted after several years of development of the freely swinging gravity meter with its horizontal accelerometers (HAMs) which made a continuous correction for the horizontal acceleration. The earlier instrument made usable measurements under smooth to moderate sea conditions. When good gyroscopes at moderate cost became available, in about 1965, from companies working for the space and military agencies, the way was opened for the development of gyro-stabilised platforms and the modern gravity meter mounted on them.

The proton precession magnetometer was invented about 1954 and rather quickly developed into a practical instrument for airborne and shipborne use. In the shipborne application, the instrument is towed at the end of a relatively long cable (approximately 150 m) to be free from the magnetic effects of the ship.

Units

In scientific and engineering literature, there is no generally accepted unit for acceleration. However, in geophysical literature that unit is called the gal (cm/s), after Galileo. In geophysics the milligal (mGal) is the usual unit and is 0·001 gal. On many gravity maps, especially of older surveys in the US, the 'gravity unit' 0·1 mGal or 10^{-4} gal is used.

In magnetics, the common unit† for magnetic field strength is the oersted and in magnetic prospecting the gamma or 10^{-5} oersted is used as a convenient unit in contouring and in magnetic calculations.

Costs and Speed

Gravity and magnetic surveys at sea are generally made in one of the two methods; i.e. (1) as an independent operation or (2) as added instrumentation accompanying a seismograph operation. The advantage of the first operation is that it is inherently about twice as fast with no significant loss of detail. There may be a further gain as the gravity–magnetic equipment is somewhat less complicated and not as

† Recently, the Committee on Units and Nomenclature of the European Association of Exploration Geophysicists recommended the adoption of the International System (SI) in geophysics. In this system, the unit of field strength is the tesla, which is 10^{-9} oersted. The recommendation of the committee has not been generally adopted.

subject to breakdown as the seismic instrument which necessarily shuts down the operation.

The second method is considerably cheaper since the cost of the ship and the navigation systems are already paid for by the seismograph operation. The necessary instruments, the gravity meter and the magnetometer and their rather complicated recording equipment, are simply added in space, preferably near the centre of the ship, which is usually available. The necessary operators, usually only two men, are simply added to a crew of some fifteen to twenty men. The cable noise usually limits boat speed to about 5 knots, whereas unencumbered, both gravity and magnetics can easily make good measurements at at least twice that speed.

The total cost (1978) of a gravity–magnetic ship and crew was around $175 000 to $200 000 per month, including the ship, navigation, gravity and magnetic equipment. This will include a technical crew of three or four men to operate the gravity and magnetic equipment, the navigation equipment, the fathometer and the recording equipment. Such a ship will produce records at a speed of 10 to 12 knots and a net output of about 4500 to 7000 km of line per month including returns to base for refuelling, restocking of food and water, crew change, etc.

The cost of gravity–magnetic equipment added to a seismic equipment is around $17 000 to $23 000 per month. This will include the gravity meter and magnetometer and possibly the fathometer although the ship's fathometer may be adequate, especially if it has or can be equipped with digital recording. Such a ship will produce about 1600 km of line per month.

GRAVITY MEASUREMENTS AT SEA

General Characteristics of Motions at Sea

Gravity is a form of acceleration and there is no physical principle by which the acceleration of motion and the acceleration of gravity can be separated. Therefore, any instrument which measures gravity at sea necessarily operates in the acceleration field imposed by the motion of the sea and the response to it of the ship on which the instrument is mounted.

The most serious motion is that due to 'heave' or the up-and-down motion of the ship. This has a period of 3–30 s with a commonly occurring period of around 10 s. To estimate the magnitude of the resulting acceleration, assume simple harmonic motion, for which

$$a = r\omega^2 = r\left(\frac{2\Pi}{T}\right)^2 \approx r \times \frac{40}{T^2}$$

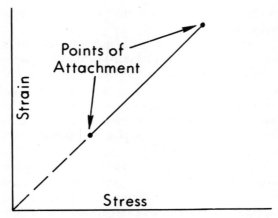

Fig. 1. Principle of zero-length spring.

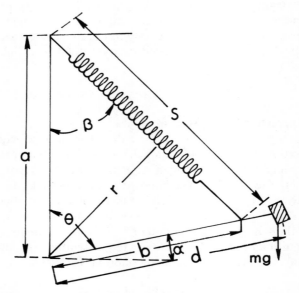

Fig. 2. Definitions for analysis of zero-length spring.

where a is the acceleration in gals, r is the amplitude of the motion in cm and T its period in s.

For a specific case, assume an amplitude of 200 cm and a period of 10 s. The acceleration is then,

$$a = \frac{200 \times 40}{100} = 80\,000 \text{ mGal} = 0{\cdot}08 \text{ of total gravity}$$

Such an acceleration represents a very severe environment in which to make gravity measurements accurate to 1 mGal. It is routinely accomplished through heavy damping in the instrument and extensive filtering in the instrumental accessories and in the data processing.

The magnetometer is normally not affected by the motions; it is subject to the magnetic effects of the ship and has to be towed behind on a long cable, usually 100–150 ft, to be far enough away so that variable magnetic effects are negligible.

The La Coste and Romberg Stabilised Platform Gravity Meter

Since the La Coste and Romberg instrument[1] is used much more than any other in commercial measurements at sea, it will be described first in some detail. The instrument is an adaptation of the land instrument which has been used for many years, and by stiffening certain members of the moving system and introducing very heavy damping, it is made operable in high acceleration fields. A system of filtering and digital recording is used, and a strip chart record is also made of certain outputs of the instrument.

The zero-length spring is basic to the design of the La Coste and Romberg gravity meter. This spring is wound in a pre-stressed condition (like an ordinary screen door spring) so that it takes a certain force before the spring windings begin to separate. It is made so that if it were capable of collapsing to zero-length, the force would be zero, as shown in Fig. 1.

The elements of the spring and its connections are shown by Fig. 2. The gravitational torque is

$$T_g = mgd\sin\theta = mgd\cos\alpha$$

The torque from the spring is

$$T_s = Ksr$$

and

$$s = \frac{b\cos\alpha}{\sin\beta}$$

$$r = a\sin\beta$$

From these relations

$$T_s = K \frac{b \cos \alpha}{\sin \beta} a \sin \beta = K b a \cos \alpha$$

When the meter is balanced,

$$T_g - T_s = 0$$
$$mgd \cos \alpha = K b a \cos \alpha$$
$$mgd = K b a$$

This shows that the instrument is insensitive to the angles θ, α and β. It may be in equilibrium over a small range of angles; there is no restoring force and theoretically it can be adjusted to infinite period and infinite sensitivity. In practice the period is usually adjusted to about 17 s in land gravity meters and to about 3 min in highly damped shipborne gravity meters.

The zero-length spring gravity meter has had some modifications over the many years since it was first used for a land meter. However, the fundamental principle has not been modified. In the modern instrument as developed for measurement of gravity at sea, the most important change has been the introduction of extremely heavy damping along with the stiffening of certain parts of the instrument to reduce cross-coupling and other undesirable effects.

Introduction of very heavy damping, together with the low restoring force, changes the basic measurement from one of displacement to one of velocity. This can be illustrated in terms of the differential equation (see reference 1, p. 483) as in Fig. 3(a).

$$b\ddot{B} + f\dot{B} + KB - cS = g + a$$

where B is the displacement of the beam, $\dot{B}$, and $\ddot{B}$ are its first and second time derivatives, cS is the combined force per unit mass exerted by the spring and beam to oppose the pull of gravity on the beam, g is gravity to be measured, and a is vertical acceleration of the meter case.

For a simplified but close first approximation, assume that the coefficients b, f and K are constant. Making the damping very large means that the value of the coefficient, f, is very large. Making the restoring force very small (a long period) means giving a low value to the coefficient, K. Because the damping is very heavy, the relative motion is extremely slow and acceleration forces are small so that the value of the term b is small. Therefore, if the spring tension S is kept constant the principal result of a change in gravity is a change in $\dot{B}$ which is the slope of the curve of position of the moving system. The physical meaning of the system is that a change

GRAVITY AND MAGNETIC SURVEYS AT SEA

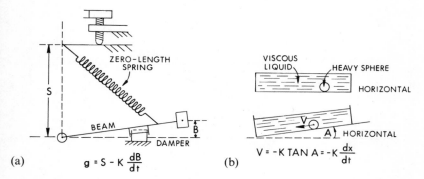

FIG. 3. (a) Zero-length spring with heavy damping; (b) analogue.

of gravity produces a rate of change of the moving system, and the slope of the curve, not the displacement itself, is the measure of gravity.

An analogy may be thought of as a flat dish filled with a heavy viscous liquid (such as honey) and containing a ball bearing (Fig. 3(b)). If the dish is perfectly level, the ball will not move. If the dish is tilted, the ball will move immediately at a rate proportional to the tilt angle A. In the gravity meter, the beam begins to move and the corresponding rate of change is determined by the unbalance between the spring tension and gravity,

$$g = cS - \frac{K\,dB}{dt} = cS - K\dot{B}$$

The equation shows that the gravity reading is the sum of two terms, namely the spring tension and the beam slope. In practice the spring tension is adjusted automatically, as shown in Fig. 5, by a slow servo to keep the beam slope small so that it can be measured accurately. The spring tension therefore accounts for most of the slow gravity changes and the beam slope accounts for most of the fast changes.

An example will help to show how the system behaves when the ship passes over an area where there is a gravity anomaly Δg as shown in Fig. 4. We will neglect the motion of the slow servo that adjusts the spring tension and assume the spring tension has a constant value equal to the value of gravity at the start of the figure. Thus at the start, the curve of displacement B versus time will have a slope of zero. Then the curve will rise at an increasing rate until it reaches the peak of the anomaly, then continue to rise at a decreasing rate and finally end with zero rate again but at a higher level.

The position of the beam is determined by photo cells and internal electronic circuitry. The spring tension screw, see 1 in Fig. 5, is balanced by a

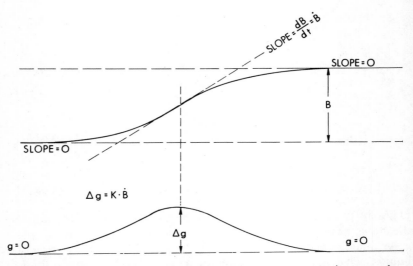

Fig. 4. Response of heavily damped moving system to gravity anomaly.

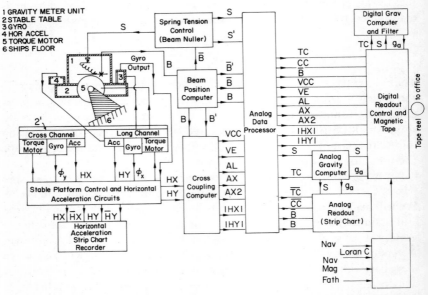

Fig. 5. Stable platform gravity meter and associated control and information circuits.

feedback circuit so that the beam is kept in a near-balanced position at all times. The table, 2 of Fig. 5, on which the entire assembly is mounted is controlled by two gyroscopes (one of which is shown at 3). These gyroscopes control two motors (one of which is shown at 5), which keep the table stabilised in space by counteracting motions of the deck of the ship. Since the table must be kept horizontal as well as stabilised, the gyroscopes are coupled through electronic circuitry to horizontal accelerometers 4 (which amount to levels). Any offlevelling of the table causes the accelerometers to slowly precess the gyros to restore the table to level. It is quite impressive to watch the performance of the gyro platform in rough seas when the boat is pitching and rolling; the torque motors are continually turning to counter the motion of the ship and the levels on the platform show an almost constant near-level condition.

The entire process is recorded on the graphical record made on the ship, as well as digitally. There are also several other recordings corresponding to the navigation and to various processing calculations which are carried out in the reduction of the data. These will be taken up later on p. 226.

The gravity meter on the stabilised platform is subject to horizontal as well as vertical forces due to the motion of the ship. These forces can interact with each other to give small effects known as cross-coupling if the gravity meter is not as stiff or as linear as it should be. It is possible to anticipate these effects, to a large extent, by certain procedures carried out during manufacture and to make allowance for them in the operation of the meter through a cross-coupling computer, as indicated in Fig. 5. The cross-coupling remains one of the most troublesome corrections made to the La Coste and Romberg meter. Means to reduce these corrections, discussed in the next section, are under development and show promise of considerable improvement in operations.

The Straight-Line La Coste and Romberg Meter

Recently a modification of the La Coste and Romberg gravity meter has been made in which the centre of gravity of the moving mass moves in a straight line rather than in the arc of a circle. This straight-line motion makes the new design insensitive to accelerations normal to the straight line, thereby eliminating any cross-coupling effects inherent in the design.

Figure 6 shows a modification of the new design which is well adapted in theory although it is not likely to be used in practice. The modification has three-fold symmetry. The figure shows the mass as a triangular prism which moves in a straight line about its symmetry axis. The figure also shows three identical spring suspensions on the three symmetrical sides of the prism.

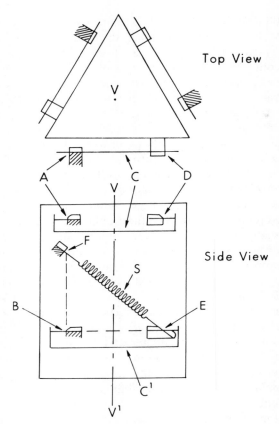

FIG. 6. Straight-line suspension.

Each of the spring suspensions behaves in the same way as the single zero-length spring suspension used in the standard shipborne gravity meter. The zero-length spring suspensions give a high displacement sensitivity.

The side view of Fig. 6 shows one of the zero-length springs. The spring is fixed to the supporting frame at F and the movable prism at E. It counteracts $\frac{1}{3}$ of the gravitational force on the prism and exerts a clockwise torque on the prism about its axis of symmetry VV^1. This clockwise torque is balanced by two compression links C and C^1. These links are attached to a supporting frame at A and B and to the prism at D and E by fine wires at their ends. The points of attachment A, F and B are in a vertical line, as are also the points D and E.

Let us now consider what happens when the prism is moved vertically downwards. The compression links are tilted downwards on their right sides. This tilt reduces the horizontal components of their length, and hence moves the right edge of the prism clockwise about its symmetry axis. Since there are similar clockwise rotations of the other two vertical edges of the prism, there is a slight rotation of the prism about its symmetry axis when the prism is displaced vertically. However, if the centre of gravity of the prism is on its symmetry axis, its only motion is along the vertical line VV^1. The suspension is then unresponsive to horizontal accelerations for all vertical positions of the prism; this is the condition for no cross-coupling errors from interacting accelerations.

As previously mentioned, the modification of the straight-line gravity meter suspension just described probably will not be used in practice. First the prism shaped mass is not as well adapted to damping as a cylindrical mass, and secondly the design can be simplified by using only two springs instead of three.

Other Instruments for Measuring Gravity at Sea

Although the La Coste and Romberg instrument has been used much more extensively than any of the others, there are few gravity meters which have not been used to some extent in making measurements at sea.

The Graf SS2 Meter[2]

This meter has a horizontal beam supported by two coiled springs so that it has a relatively long period and becomes quite sensitive. The horizontal beam is held in the gap of a permanent magnet which produces electromagnetic damping. The beam is held by ligaments in very close air gaps to make the damping quite effective. It is mounted on a stabilised platform of the special design by the makers of the instrument. It has been used to some extent in Europe and by the Lamont Laboratory, who have made considerable improvements in its operation.

The Graf SS3 Sea Gravity Meter[3a]

The design of this meter is entirely different; it uses a main spring within a vertical tube in a completely symmetrical arrangement. The tube is constrained by five ligaments and allowed to move only along its vertical axis so as to largely eliminate the effect of cross-coupling. The tube is the principal weight which is supported by the coiled spring. The small residual force is supplied by an electrical system which controls the current in a balancing coil through an oscillator, amplifier and phase-sensitive rectifier. Thus, the reaction of this coil with a permanent magnet measures the change of gravity and vertical acceleration due to motion. The separation of the motional component is based on time filtering in the output.

The Ambac Industries Instrument[4]
This instrument is based on a double vibrating string with a mass supported between the two strings. As gravity increases, one string increases in frequency and the other decreases and the difference in frequency is a measure of the change of gravity. The strings are in the field of permanent magnets and are driven at their resonant frequency by suitable electric circuits. Cross-coupling effects are eliminated by cross support ligaments which keep the centre of mass of the two weights and the entire assembly aligned in the vertical axis.

Tests[5] indicate a reliability of 0·7–1 mGal in moderately rough seas. Results of similar reliability were obtained in tests by the US Navy Applied Science Laboratory, and by the Bedford Institute.

The Bell Accelerometer[3b]
This meter uses a coil in the permanent field of a pair of magnets. The weight of the coil and the accompanying 'proof mass' is balanced by a constant current in the coil maintained very accurately by the external electronic system which takes the place of the main spring in most gravity meters. Small changes in this current automatically compensate for the small changes in weight resulting from changes in gravity and motional acceleration. These are measured, recorded and filtered to separate the desired changes in gravity from motional accelerations, by a small computer which operates on the digital signal.

Corrections

Eötvös Correction
The Eötvös correction compensates for the apparent gravity effect due to motion over the surface of a rotating earth. This motion modifies the acceleration due to earth rotation, and as the effect can be quite large it must be calculated and removed.

At a point on the equator the centrifugal outward effect due to rotation is about 3500 mGal and decreases to zero at the poles. This acceleration is perpendicular to the earth's axis and is $R\omega^2 \cos\varphi$. The component perpendicular to the earth's surface is $R\omega^2 \cos^2\varphi$, as shown in Fig. 7. Any

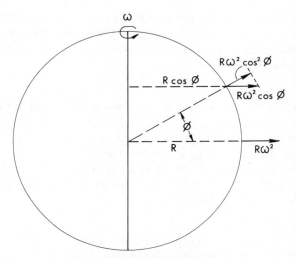

FIG. 7. The Eötvös correction.

east or west motion produces an increment, da, in this acceleration which, by differentiation, is,

$$da = 2R\omega \cos^2 \varphi \, d\omega$$

The change due to the east component of the ship's motion, V_E, is obtained from

$$V_E = R\cos\varphi \, d\omega \quad \text{and} \quad d\omega = \frac{V_E}{R\cos\varphi}$$

and the Eötvös effect, da, is

$$E = da = 2V_E \omega \cos\varphi = 2V\omega \cos\varphi \sin\alpha$$

where V is the velocity of the ship and α is the heading angle between the direction of motion and astronomic north.

There is a second correction, V^2/R, which is the simple outward effect due to motion over a curved surface, independent of direction. It is usually neglected as it amounts to only 0·4 mGal for a speed of 10 knots.

To reduce the above expression to common units, we take the radius of the earth as $6·371 \times 10^8$ cm and ω as $7·2921 \times 10^{-5}$ rad/s (from $\omega = 2\Pi/T$, where T is the length of the sidereal day or 86 164 s), which gives

$$E = 14·584 \times 10^{-5} \, V \cos\varphi \sin\alpha$$

for speeds in cm/s. To change to speed in knots, 1 knot = 1.85325 km/h = 51.479 cm/s and $E = (14.584 \times 10^{-5})\, 51.479 = 7.508 V \cos\varphi \sin\alpha$ mGal/knot. The correction is positive when moving eastward and negative when moving westward.

Latitude Correction

The latitude correction is the same as on land and it is usually taken directly from tables of gravity as a function of latitude. This is usually calculated by the computer, either on the ship or later in the office, with the latitude of observation from the navigation. When the value of gravity, corrected for latitude, is subtracted from the value given in the tables, the final value is reduced to the deviation from the normal field of the geoid.† The value may be either positive or negative.

The Meter Zero

A desirable practice is to keep a long-time record of all 'still readings' of each meter, made when the ship is at the dock. If an absolute gravity value is available at the dock, tied into the international net, a 'meter zero' can be determined. This is simply the value of gravity, which, if it could be attained, would reduce the meter reading to zero, or

$$G_0 = G_a - G_s$$

where G_s is the still reading and G_a is the absolute gravity at the dock and G_0 is the meter zero.‡ A graph of still readings against time will give a picture of the long time performance of the meter.

Elevation or Water Depth Correction

In gravity measurements at sea there are, of course, no elevation corrections since all the measurements are made at sea level. Theoretically, there should be tidal corrections but these are usually ignored since there is no practical way of making them and they are usually adjusted out in the closure corrections. They are seldom recognisable since they would rarely amount to 1 mGal. Corrections for the tide should be made when tying into reference stations at a dock where the rise and fall of the water level is

---

† The value of the first constant of the gravity formula has now been reduced by 14 mGal because of recent absolute gravity measurements by a falling body.[3c]
‡ The principle of the still reading can be applied to any meter which has a continuous screw for world-wide readings. The La Coste and Romberg shipborne meter does not have a buoyancy compensating cell and is therefore sensitive to leaks and also will show a discontinuity in the still reading curve if the case is opened for any reason.

readily recognisable. Usually this is done by simply adjusting later still readings at the dock to the first reading made which will have the usual 'free air' correction (0·3086 mGal/m or 0·094 06 mGal/ft) for the difference in elevation of the dock and the meter.

Water depth (or 'Bouguer')† corrections are usually made from fathometer readings. These reduce the readings to what they would be if the rock of the density of that at the sea floor extended to the surface. The correction is usually made directly from the fathometer reading, either on the ship or later in the office.

If σ_r is the density of rock of the bottom and σ_w is the density of sea water, the correction is:

$$f = 0·012\,76\,(\sigma_r - 1·03) \text{ for depth in ft}$$
$$f = 0·076\,56\,(\sigma_r - 1·03) \text{ for depth in fathoms}$$
$$f = 0·041\,85\,(\sigma_r - 1·03) \text{ for depth in m}$$

Terrain Correction

The terrain correction is quite different from that on land. When the depth of water is determined by the fathometer within or near an area of topographic relief, the surface may be higher or lower in areas nearby. Consider an area near a submarine cliff. As the ship approaches the point P, in Fig. 8, from the left at position 1, the fathometer does not see the lower topography to the right so the Bouguer effect, calculated as if the sea floor were level, is too low and the correction is positive. On passing point P, at position 2, the fathometer does not sense the higher topography to the left, the Bouguer effect calculated is too high and the correction is negative. Note that the Bouguer correction would have a large discontinuity in crossing the cliff, but in the case illustrated, where the sea floor has a vertical cliff of 1 km relief and the densities are as shown, the calculated terrain correction is 35 mGal on either side of the cliff or a total of 70 mGal. If the cliff is sloping, the effect will not have the sharp peaks but will be rounded off as shown approximately by the dashed line. In one example, off the Californian coast, where there was a relief of several thousand metres, the terrain corrections were over ± 26 mGal.

† The water depth correction and the Bouguer correction as used for land gravity are not quite the same. The measurement on land is corrected for readily determined density which is the local density of the surface rock. At sea, the water depth correction is a first step in the interpretation. This leads to the almost universal phenomenon of a rise in the Bouguer gravity in crossing the continental shelf because the change to the deep ocean is generally compensated, the free air anomaly is near zero, on the average, and the Bouguer anomaly shows a large positive value.[3d]

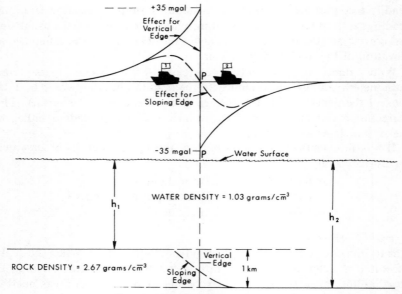

Fig. 8. Terrain correction at sea.

MAGNETIC MEASUREMENTS AT SEA

Introduction

Magnetic measurements are necessarily made in the magnetic field of the earth. This consists of three parts:

(1) A comparatively large part which is due to the earth as a whole and its internal structure;
(2) A comparatively small and variable part which changes with the time of day in a somewhat regular manner and which also has a more or less erratic change superimposed upon it. This may vary up to 100 gamma and may, at times of magnetic storms, become so large and erratic as to make magnetic operations useless;
(3) A part which is the objective of the magnetic field measurement and varies from a fraction of a gamma to several thousand gamma but commonly is less than 100 gamma. This part is due mostly to the 'basement' igneous or metamorphic rocks. The determination of depth to basement is a principal objective of magnetic measurements.

The magnetometer is towed some distance behind the ship to keep it from being influenced by variation of the magnetic field introduced by iron parts or electric currents. The distance will depend on the magnetic characteristics of the ship, but it is usually 100–150 m.

Magnetic Units

It is necessary to consider briefly the elements of the magnetic field and their definition before discussing the instruments for the measurement of the field. The unit commonly used in magnetic prospecting is the gamma, commonly represented by γ; one $\gamma = 10^{-5}$ oersted, the cgs unit of magnetic field strength.† The magnetic field is a vector in space which may be defined by three of its components. Let us consider it represented by the diagonal of the box, Fig. 9. Then,

X, Y and **V** are the north, east and vertical components,
T is the total component,
H is the total horizontal component,
d is the angle between the true north and the total horizontal component (the ordinary magnetic declination),
i is the angle of inclination.

The following relations exist between these quantities:

$$\mathbf{T} = \mathbf{H}/\cos i; \; \mathbf{T}^2 = \mathbf{H}^2 + \mathbf{V}^2; \; \mathbf{T}^2 = \mathbf{X}^2 + \mathbf{Y}^2 + \mathbf{V}^2$$

$$\mathbf{X} = \mathbf{H}\cos d; \; \mathbf{Y} = \mathbf{H}\sin d; \; \mathbf{V}/\mathbf{H} = \tan i$$

In magnetic measurements at sea, it is always the total intensity of the magnetic field which is measured. Thus, the field can vary from $+60\,000\,\gamma$ through zero to $-66\,000\,\gamma$ depending on where on the surface of the ocean the measurements are made. The measured field is reduced by removing the magnetic field as mapped for the whole earth by the International Geophysical Reference Field (IGRF).

The Proton Precession Magnetometer

The instrument most commonly used for magnetic measurement at sea is the proton precession magnetometer. This has the advantage of measuring in absolute units so that the value of the IGRF can be subtracted (by a computer operation) to give the anomaly directly. The proton precession magnetometer operates on the precession of the atomic nucleus, usually of hydrogen, from a bottle of water.

† See footnote, p. 206.

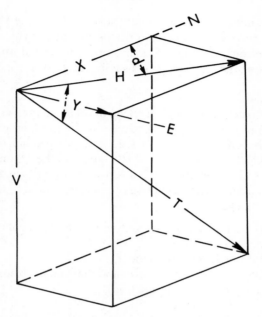

Fig. 9. Components of magnetic field.

An atomic nucleus has the property that when it is magnetised in a strong magnetic field and then the field is suddenly reduced to zero, the spinning nucleus precesses at a rate which is proportional to the ambient magnetic field; the phenomenon known as the Larmor precession. The protons normally have their spin in random directions. They can be lined up with their axes all in the same direction by a strong magnetic field, behaving like a group of tiny tops all spinning together. If, now, the magnetising field is removed, they precess together at a frequency controlled by the strength of the ambient field. This frequency of precession is

$$f = \gamma \mathbf{T}/2\Pi$$

and

$$\mathbf{T} = 2\Pi f/\lambda$$

where $\mathbf{T}$ = the magnitude of the ambient field and λ is the gyromagnetic ratio.

The gyromagnetic ratio, λ, is one of the fundamental physical constants of nature; it is known to high precision and is independent of the surroundings or chemical composition. Its value is:

$$\lambda = 2.675\,13 \times 10^4 \text{ (oersted-second)}^{-1}$$
$$= 0.267\,513 \ (\gamma\text{-s})^{-1}$$

The total field (**T**) in γ is:

$$\mathbf{T} = \frac{2\pi}{0.267\,513} \times f = 23.4874 f$$

Thus, if the frequency of precession can be measured, the field strength is determined. If the earth's field is, for example, $55\,000\,\gamma$ (northern Atlantic coast)

$$f = \frac{\mathbf{T}}{23.4874} = \frac{55\,000}{23.4874} \approx 2300 \text{ Hz}$$

The Varian instrument, generally used in magnetic measurements at sea, uses a high frequency counting circuit (of the order of 100 kHz) to measure f. An electronic gate opens the counting circuit and then closes it when a certain number of whole cycles, at the frequency f, has passed, thus timing that number of whole cycles. This gives a quantity which is the inverse of f (the greater the value of f, the smaller the count) which is then inverted electronically to give the true f. Thus, the proton precession magnetometer gives an output which is directly proportional to the total field strength in γ.

Because of the magnetise-and-then-count cycle, the operation of the magnetometer is inherently intermittent with a frequency of about 1 Hz. At this frequency, the spacing between points over which the field is measured is only about 5 m for boat speeds of 10 knots.

Single and Double Recording

In cases where the diurnal effect is troublesome, it may be desirable to use a double recording of the magnetic data which is useful in recognising diurnal effects. If one instrument is towed at, for example, 100 m behind the ship and the other at 200 m behind, diurnal effects will reach both instruments at the same time. A true anomaly from a fixed disturbance in the earth will be apparent as it will arrive at the farther detector at a later time than at the nearer one and the difference in time will be determined by the ship's speed.

NAVIGATION

Introduction

Navigation is most important for measurement of gravity at sea, not only for the general purpose of mapping the results but particularly for determination of the Eötvös effect. It is highly desirable that the ship's course be as nearly as possible on a straight line and be recorded properly at all times.

Systems of Navigation
Fixed Radio Patterns

All the usual methods of navigation are used. Several systems depend upon a fixed network of radio waves in space that are laid down by interfering patterns of constant radio frequency waves. The frequencies generally used are just above the ordinary broadcast range. These provide 'lanes' at a spacing of half wave length in the line joining two principal stations and at wider spacing, reaching several times the minimum, at points far off that line. Another set of similar 'lanes' is formed by a second pair of stations (the central one is in common with one of the first pair) so that the two sets of lanes, commonly designated red and green, cover the area. By ascertaining the location within the lanes as well as the count of the whole number of lanes, the position is determined, commonly within 3 m or less. The position in ordinary rectangular map coordinates is then determined by an electronic computer.

Such systems are Raydist, Lorac, Toran and Decca, which are all designed to operate within 150–200 km of the shore and depend on signals from fixed and generally quite permanent stations. They can operate with any number of boats within their pattern.

Another system known as 'range–range' operates from two fixed shore stations plus a station on the boat. It has a more advantageous pattern in that the 'lanes' are all circular in form and equidistant from one another. The two distances from the fixed shore stations give the location. It has a disadvantage in that only a very limited number of boats, usually only one or two, can operate in the same pattern.

Satellite Navigation

Satellite navigation can be used when there is some way to fill in between the fixes by satellite which may be from one to two or three hours apart. Such fill-in may be made in shallow water (less than 330 m deep) by sonar Doppler navigation which depends on signals reflected from the bottom. The system depends on transmitting one set of signals forward and another backward and a second pair to the right and left. The difference in frequency of one pair of signals is the measure of the velocity. The two pairs of signals give the component of velocity in the direction of the ship and the component perpendicular to that direction. By integrating these with time, the two components of distance are determined. The accuracy is about 1 % of the distance travelled. Thus, if the position at one to three hour intervals is determined from satellites and the intermediate position is interpolated by sonar Doppler adjusted to the satellite fixes, the ship's course can be quite well determined.

Another system is the use of Loran C for the intermediate fill-in of satellite fixes. Loran C is a system which uses relatively long radio waves from a pattern of widely spaced radio stations at fixed locations along the coasts of the US and other countries. These stations determine a fixed pattern of interfering waves which can be used for locations between satellite fixes.

Third Axis Gyro

A recent development[6,7] is the use of a third gyro with the gravity meter, to stabilise it about the vertical axis as well as about the two horizontal axes. The components of the ship's speed are computed by post-processing data from the three gyros and the two accelerometers to give the Eötvös correction. Since inertial navigation is subject to long-time errors resulting from changes in gyro drifts or accelerometer errors, the inertial results are corrected either from satellite fixes or comparison with Loran C measurements. Loran C works well with inertial navigation because it gives an Eötvös correction that is accurate over long times where the inertial Eötvös is inaccurate and vice versa. A test of an inertial system recently made shows promise of being able to make navigation sufficiently accurate for the Eötvös corrections without any shore based control.

DATA PROCESSING

Modern shipborne gravity and magnetic operations depend to a very large extent upon computer processing, both in the recording of the data and in the subsequent calculation and reduction to a final map.

On the Ship

The various elemental quantities are observed as electrical outputs which are then filtered, combined, returned as controls or recorded on strip charts. This is a rather complex process as illustrated in Fig. 5 (page 212), which is modified slightly from one in the La Coste and Romberg operator's manual.

The various interactions between the elements of the meter and stable platform, on the one hand, and the controls exercised by the various computer and data processing elements, on the other, are quite complicated as indicated by the figure. The main features are as follows:

(1) The cross gyro 3 controls the cross torque motor 5 to stabilise the platform 2 about the cross horizontal axis. The similar stabilisation about the long horizontal axis is not shown in the figure;

(2) The cross accelerometer 4 precesses the cross gyro to level the platform about the cross axis. Similar levelling about the long axis is not shown;

(3) The beam position computer and spring tension control units use the gravity meter beam position B as a signal to control the spring tension S to slowly null B.

The two strip chart recorders make graphical records; one shows the horizontal acceleration, for general indication of the roughness of the sea and the operation of the equipment; the other gives the output of the gravity measuring equipment on four traces, i.e. the average beam position B, the spring tension S, and the average total correction TC which together with S give the gravity ga. These traces can be used to save the data should there be a failure in recording on magnetic tape.

The navigation data entered into the system may be satellite data together with a means of interpolating between fixes, which may be sonar Doppler, Loran C or some other system. The magnetic data may be entered here. It is the digital output of the Varian precession magnetometer in absolute units. The fathometer data may also be entered here if the ship is equipped with digital recording of the depth. Otherwise, it is left until later when it is digitised on cards and entered onto the proper tape. All the various kinds of data are recorded on a single tape, which together with the necessary records of the line numbers and times of beginning and ending, properly correlated with the reel number, complete the record.

If the ship is used also for seismic operations, some of the data may be common to the two systems and will come from the seismic records. This is particularly true of the navigation data and to a lesser extent the water depth. If the ship is not equipped with a suitable fathometer or the depth is beyond its range, the depth may be obtained from the first arrival times of the seismic records.

In the Office

The work in the office begins, of course, with the tape from the ship, at stage 1 of Fig. 10 (the system illustrated is taken from the operations of Tidelands Geophysical Co.). At stage 2 the navigation data are separated

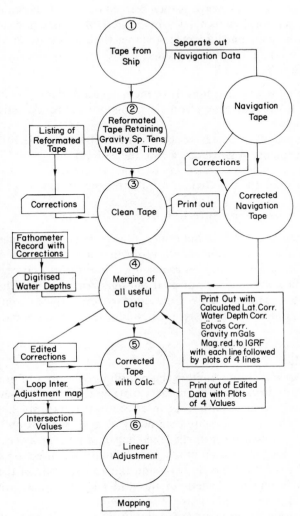

Fig. 10. Steps in processing information from ship.

out and the remainder is reformated, retaining only those components which are necessary. This tape is then printed out; it will contain obvious errors for which corrections must be made on IBM cards. The clean tape, at stage 3, is printed out and inspected, and then the digitised water depths and the navigation data are put in at stage 4 and the various calculations are made. These will include:

(1) The latitude correction from the navigation data;
(2) The Eötvös correction from the navigation data and possibly a gyro compass;
(3) The gravity reading reduced to mGal by the table (or, in later instruments, a constant) giving the calibration;
(4) The magnetic reduction to the International Geophysical Reference Field (IGRF) from the magnetometer and navigation data;
(5) The water depth correction from the fathometer data.

This is printed out and four quantities, free air gravity, Bouguer gravity, Eötvös correction and water depth correction are plotted for each line following the numerical data for that line. The obvious corrections are made and the corrected tape printed out again at stage 5 with the four curves plotted again. Finally, the intersection values from a loop intersection adjustment map are linearly adjusted at stage 6 and the final map is made.

The Final Map

The final map will always contain the position of the lines of gravity and magnetic data with values a few minutes apart on the lines. The lines should always be laid out in a pattern of intersecting traverses so they can be adjusted for closure. These are commonly in grids which are 3 × 8 km, 6 × 12 km, 3 × 10 km and so forth depending on what degree of coverage is required. The long dimension of the grid is usually in a direction along the strike and the short dimension perpendicular to the strike of the general area, if that is known. Sometimes the grid may be square with equal dimensions in the two directions.

The linear adjustment is made for closure around the loops of the net. The quality of the survey may be determined by the size of these adjustments; if the adjustment is 1 mGal or less and 10γ or less over 7·5–12 km loops, the quality is very good. It may run all the way to adjustments of several mGal and many γ over loops that are much larger and irregular in form, made for reconnaissance surveys.

Other Data Processing Systems

There are other systems used for computing of results of gravity and magnetic surveys. Most of these are not published but all will contain the elements described above. Some will do more on the ship than described here, particularly if the gravity and magnetic equipment are added to a seismic ship.

A ship with very sophisticated computer equipment on board is the Gulfrex operated by the Gulf Oil Co. described by Darby et al.[8] (A later ship, Gulf's *Hollis Hedberg* is now in operation and carries a similar installation.) While the description is largely concerned with that required for the seismic equipment, the ship also carries gravity and magnetic instruments and the description includes details of the data processing which gives real-time gravity and magnetic records processed with the usual reductions and presented as plots properly placed on the seismic records.

The equipment used includes,

(1) four aqua-pulse seismic sources,
(2) a 24-track seismic cable,
(3) a La Coste and Romberg gravity meter,
(4) a Varian proton precession magnetometer,
(5) a fathometer,
(6) a gyrocompass,
(7) shoran,
(8) sonar Doppler,
(9) a velocimeter,
(10) a satellite receiver,
(11) a sal log,
(12) a master clock.

Not all of these are used at all times and not all are required for the gravity and magnetic data. Those that are used, besides the gravity meter and magnetometer, are the fathometer for water depth data, and some combination of the gyro compass, velocimeter, shoran and satellite receiver for navigation (depending on where in the world the ship is working), for the latitude, and Eötvös correction, and also for reduction of the magnetic data for the IGRF.

The real-time gravity and magnetic records, with the usual corrections for latitude, Eötvös effect, water depth and reduction to the IGRF, are produced and superimposed on the single trace seismic records. Thus the seismic record and the corresponding gravity and magnetic record can be examined together immediately. If there is any detail that requires repeating or filling-in with one or more cross lines or parallel lines, the program can be undertaken immediately without waiting to process the data.

INTERPRETATION OF GRAVITY AND MAGNETICS AT SEA

Introduction

Fundamentally, the same principles apply to the interpretation of gravity and magnetic data, whether obtained on land or at sea. However, there are certain differences which are imposed by differences in the field operations.

Usually, at sea, the data are recorded at the same time, on the same lines, and at the same elevation (sea level). On land, the surveys may be made at different times and at widely different elevations, if airborne magnetics are used. At sea, lines are frequently relatively widely spaced so that two-dimensional approximations and interpretation methods may have to be used. Line spacing may be from 1·5–15 km or more with occasional nearly isolated lines. Generally, marine surveys are less accurate than on land because of the use of filters in both the original recording and in the processing of the data. Contouring is generally at 1 mGal intervals for gravity and 10γ intervals for magnetics.

When gravity and magnetics accompany a seismic line, they generally give additional clues to the resolution of the data. For instance, magnetics may give an indication of basement depth when it is beyond the depth where the seismic reflections are clearly meaningful. The gravity may show large basement structure or intrusions which are not definitely indicated by the seismic reflections. In general, the more different kinds of data available the more certain is the resolution of the problems of a given area.

Very sharp changes in gravity and magnetics that would be measured by surface measurements on land, may, over deep water, be smoothed out because the instrument is necessarily at a distance from any possible source, depending on the depth of the water. Sharp effects also may be smoothed out to some extent by the filtering in the meter itself or in the data processing.

Lines and Closed Loops

Gravity and magnetic measurements at sea are nearly always constructed as profiles of intersecting lines. The individual lines may be from a few km to as much as 150 km long. The net of intersecting lines can be checked and adjusted for closure of individual loops. If the net is fairly close, the results can be contoured and the interpretation made in much the same manner as on land. As the net becomes wider, interpolation between lines of control becomes more difficult. Finally, if the lines are isolated from each other or are open-ended, the interpretation will be largely by two-dimensional methods and calculations.

If the control is fairly close, as it may be if the gravity–magnetic equipment is added to a seismograph ship engaged in rather detailed operations, the profiles may be laid out in the most favourable direction. Finally, if sufficient detail is available, the map may be subjected to the same processes of 'second derivative', filtering or Fourier transforms that are used on land surveys.

In general, gravity at sea is not as precise as on land. The large corrections for Eötvös effects, which depend on accurate navigation, the filtering, and the adjustment and smoothing of the final profiles, are such that a contour interval in the final maps of less than 1 mGal is not usually justified. On the other hand, with careful attention to detail, especially navigation, very good maps can be made and 1 mGal contours can be quite reliable.

Properties of Gravity and Magnetic Fields

According to potential theory, the magnetic field is one derivative higher and therefore is sharper than the gravity field.[3e] This must be kept in mind when comparing gravity and magnetic maps made from the same lines of control. Furthermore, practically all magnetic features come from basement rocks while gravity anomalies can have their source either in density contrasts within the basement or in the overlying sedimentary section.

Magnetic maps made in lower latitudes, where the dip angle is less than about 50–60°, tend to have a low on the magnetically north side. Thus, for the expression of the same source, there may be a single gravity high, together with a magnetic high centred on the south side and a low centred on the north side of the gravity high.

Depth Determination

The determination of the depth of the basement is one of the principal and most valuable operations which can be carried out on a magnetic survey. There are many methods of making depth estimates.[3f] Most operate on profiles which may be either the profiles made along the ship's course or from the contours made on the map; the former may have greater detail and the latter may be better related to geological strike.

Usually, depth determinations are not as reliable as those made from airborne surveys, as the critical changes in slope are not so well determined.

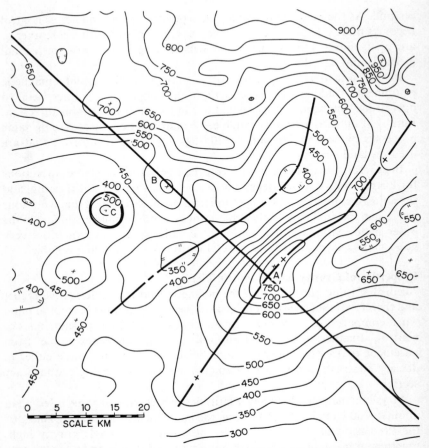

Fig. 11. Bouguer gravity map. Contour interval 5 mGal. A, B and C are locations of anomalies discussed. Trends of large anomalies indicated.

Interpretation of the Maps

The Bouguer gravity and magnetic maps presented in Figs. 11 and 12 are selected from a much larger survey. The Bouguer gravity is reduced for a density of 2·67. The total magnetic intensity has the earth's normal field (IGRF) removed.† A water depth map (not included) is also available.

† Grid values of the IGRF, 1965 are given by Environmental Science Services Administration (ESSA) Tech. Ref. C & GS 38, 1969. The values also may be furnished as a computer program.

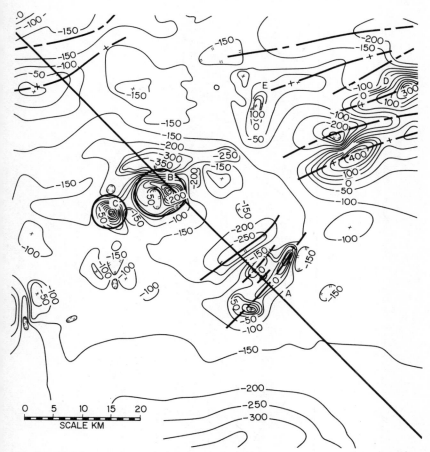

FIG. 12. Total magnetic intensity map. Contour interval 50 γ. Inclination about 50°. A to E are locations of anomalies discussed with trends of anomalies indicated.

The maps were made from a regular grid of survey lines with spacing of 4 km N–S by 10 km E–W. All loops of the survey were adjusted for closure by least squares. The profiles, Fig. 13, are drawn from Bouguer gravity, total magnetic intensity and water depth maps along the approximately NW–SE line shown on the two maps.

The gravity map is dominated by large features trending approximately N–30° E, across the maps, some of which are marked (A, B and C). The magnetic map is more complicated because of the minima shown on the

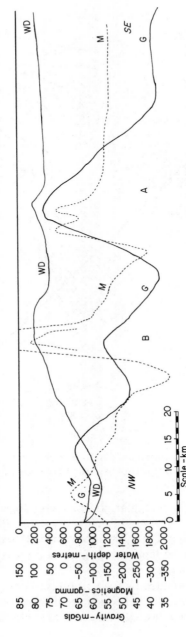

FIG. 13. Profiles of Bouguer gravity (*G*), total magnetic intensity (*M*), and water depth (*WD*), on the NW–SE line of the maps. A and B are the gravity-magnetic anomalies discussed in the text.

north side of the magnetic highs which are fairly prominent at this magnetic latitude which is about 50°N.

It must be kept in mind that, because they are higher derivatives of the potential function, the magnetic anomalies are much more sensitive to depth than the gravity anomalies. Both the large anomalies of the magnetic profile occur in areas of relatively shallow water and their tops must be at or near the bottom of the water because of their sharpness. The gravity anomalies, which may be caused by the same body, are broader, smoother and their effects are much deeper.

Analysis of the Gravity Anomalies
Taking them as two-dimensional, for the gravity at the centre of the anomaly,[3g]

$$g = \frac{2\Pi G \sigma R^2}{z} \quad \text{and} \quad z = X_{1/2}$$

where g is the amplitude of the gravity anomaly in mGal, G is the gravitational constant $= 6.67 \times 10^{-8}$, R is the radius of the cylinder, σ is the density contrast, $X_{1/2}$ is $\frac{1}{2}$ the distance between the two points where the amplitude is $\frac{1}{2}$ the maximum. Putting in numerical values for G and for units in km,

$$g = \frac{42 \sigma R^2}{z}$$

For the larger anomaly, A, the amplitude is roughly 45 mGal and the double half-width is roughly 18 km giving a depth ($z = X_{1/2}$) of 9 km. Guessing at a density contrast of 0.25, we have:

$$45 = \frac{42 \times 0.25 R^2}{9}$$

$$R^2 = \frac{45 \times 9}{42 \times 0.25} = \frac{405}{10.5} = 40$$

$$R = 6.3$$

This gives depth to top $= 9 - 6.3 = 2.7$ km.

For the smaller anomaly, B, the amplitude is roughly 60 mGal, the double half-width is roughly 11 km, giving a depth of 5·5 km and, for the same density contrast,

$$R^2 = \frac{60 \times 5\cdot5}{42 \times 0\cdot25} = \frac{330}{10\cdot5} = 31$$

$$R = 5\cdot5 \text{ km}$$

This gives depth to top $= 5\cdot5 - 5\cdot5 = 0$.

The approximation of the bodies as cylinders is only very crude but it gives general limits on their sizes. The assumption of 0·25 for the density is not as critical as it might seem as the change in calculated radius is only the square root of any change in the density.

The fact that the larger gravity anomaly has the smaller magnetic expression and vice versa can be attributed to details in the shape and variation in the magnetisation. The figures are only very approximate, but we can say that the area across the middle of the map contains large intrusive bodies, with a generally ENE–WSW trend, and that they come up to the bottom of the water.

Other Anomalies

There are several gravity and magnetic anomalies, some of which correspond rather closely and others that do not. In general, the areas of shallow depth are also areas of strong magnetic anomalies, except in the southern part of the map.

There are a few cases where the gravity and magnetic anomalies correspond roughly as they should for coming from the same source, taking into account the distortion of the magnetic anomaly on account of the relatively low magnetic latitude.

Areas A and B have been discussed in some detail. They agree qualitatively, both having negative magnetic anomalies on the north side. At area C, the agreement is quite good, with a round gravity anomaly in approximate agreement with a round magnetic anomaly with a minimum on the north side. At area D, there are strong gravity anomalies and magnetic anomalies with a minimum on the north side. These are not very near the ideal relationship, indicating that the magnetic and gravity material do not have the same boundaries. There is also a high in the water depth map with a minimum depth of about 500 m corresponding closely with the magnetic high. This suggests that igneous material comes up to the water bottom. The irregular magnetic highs in Area E correspond approximately with a minimum in water depth and with a disturbed area on the gravity map. Again, these two anomalies do not have the proper relationship for coming from the same body.

On the whole, the gravity and magnetic maps indicate that the area is one of an igneous terrain, with a fairly irregular surface. There is little indication that the sub-surface gives sufficient sedimentary section to make it an oil prospective area.

REFERENCES

1. LA COSTE, L. J. B., CLARKSON, N. and HAMILTON, G., La Coste and Romberg stabilized platform gravity meter, *Geophysics*, **32** (No. 1), p. 99, 1967.
2. GRAF, A. and SCHULZE, R., Improvements on the sea gravimeter GSS2, *J. Geophys. Res.*, **66** (No. 6), p. 1813, 1961.
3. NETTLETON, L. L. *Gravity and magnetics in oil prospecting*, a–p. 29, b–p. 45, c–p. 58, d–p. 290, e–p. 376, f–p. 394, g–p. 192, McGraw-Hill, New York, 1976.
4. WING, C. G., MIT vibrating string surface ship gravimeter, *J. Geophys. Res.*, **74**, p. 5882, 1969.
5. BOWIN, C., ALDRICH, T. C. and FOLLINSBEE, R. A., VSA gravity meter system: tests and recent developments, *J. Geophys. Res.*, **77** (No. 11), p. 2018, 1972.
6. VALLIANT, H. D. and LA COSTE, L. J. B., Theory and evaluation of the three-axis inertial platform for marine gravimeters, *Geophysics*, **41** (No. 3), p. 459, 1976.
7. LA COSTE, L. J. B., 1974 test of La Coste and Romberg inertial navigation system, *Geophysics*, **42** (No. 3), p. 594, 1977.
8. DARBY, E. K., MERCADO, E. J., ZOLL, R. M. and EMANUEL, J. R., Computer systems for real-time marine exploration, *Geophysics*, **38** (No. 2), p. 301, 1973.

Chapter 7

PULSE SHAPING METHODS

D. G. STONE

Seismograph Service Corp., Tulsa, Oklahoma, USA

SUMMARY

The extraction and correction of the distorted seismic wavelet yields many interpretive advantages. Well log comparisons, line ties, fine stratigraphic definition, detection of thin layers, and data quality are all improved by successful wavelet processing.

Estimates of the reflection coefficients, though restricted to those physically possible by seismic methods, can improve detailed interpretation. Many problems such as the effect of layer curvature are not included in the estimations described. Yet the simplified seismic display is often beneficial in making interpretive decisions.

The history of geophysical technology indicates that improvements in wavelet processing and reflectivity estimates will surely come about. Iterative methods for decomposing the seismic trace to a wavelet and spike series show considerable promise. It is clear, however, that the utility of such techniques must be used in the light of the basic assumptions made about the seismic signal. The skill of the interpreter in correlating and evaluating data will still be needed even if computations are carried out perfectly. The potential of decomposing the trace is considerable and it is not yet a mature art.

INTRODUCTION

The seismic section which best enables the interpreter to make a right decision is the object of seismic data processing. When the decision involves large structural traps the estimation of the velocity is the keystone to this objective. From velocity information the normal moveout correction is used to optimise the stacked section. Subsequently the unravelling of the structural elements by proper migration is dependent on the knowledge of interval velocities. When the decision to be made concerns small stratigraphic traps, the shape and frequency content of the recorded wavefront comes to be the controlling parameter. Thin sands, small reefs, subtle lateral variations in reflection groups, and other stratigraphic detail require better representation.

The interpreter must decide if observed changes in the appearance of reflection sequences are the result of lithological variations. If the shape of the seismic wavefront changes in time or space, as it is believed to do, a troublesome unknown is involved in the decision making. To remove this variable from the data, the shape of the waveform or recorded pulse must be known. Then, with modern signal processing techniques, the pulse can be standardised in time and space to the best shape for resolution. A general consensus is that the optimal shape for resolution of thin layers is a zero phase, i.e. symmetrical pulse, whose duration is as short as possible. If this is successfully accomplished, several interpretive benefits can be realised. The correlation of adjoining seismic lines and well log data is greatly simplified. Thin layers are detectable as peaks (which are more visible than first rise-times) to the limit of the frequency content of the seismogram. Even below this resolution limit, variations in pulse form from symmetry may be interpreted in terms of the stratigraphy. Any slow, progressive character change, however subtle, is probably an expression of lithological change.

Of course the complete recovery and correction of the pulses is an ambitious undertaking. As has always been the geophysical custom, imposing barriers in the earth filtering will not damp an enthusiastic effort to capture the elusive pulse. Even more ambitiously, a complete collapsing of the pulse to the reflection coefficients is an existing endeavour. Assumptions will be made, approximations used, but computations will be carried out precisely, with optimism that the seismogram can be made to approach closer to the true depiction of the earth. All seismic processing and interpretation is however based on this philosophy and has historically worked quite well.

The specific objectives of waveshaping can be stated as follows:

(1) Remove the variances in frequency content in time and space so that anomalies observed are lithological;

(2) Correct any time and space variant phase changes so that line ties and well log correlations can be more easily made;
(3) Optimise the waveshape for the best resolution of closely spaced reflectors.

Processing which changes the waveshape then can be evaluated in the light of these objectives.

THE REFLECTION COEFFICIENTS

Reflection seismology is the most widely used technology in the search for hydrocarbons. The principles of physics for other phenomena involving wave theory are generally applicable to seismic recording. An energy source on the surface generates a spherical wave front which is transmitted, reflected, diffracted, and refracted. Surface sampling of the returning wavefront yields data depicting the subsurface.

This returning wavefront is a function of the reflectivity of the earth layering. The reflectivity of each layer is determined by the contrast between the velocity and density of successive interfaces. The greater the contrast of the interfaces, the more of the downgoing energy is reflected back to the surface. The reflection coefficient at each layer interface is

$$R_t = \frac{\partial_t V_t - \partial_{t+1} V_{t+1}}{\partial_t V_t + \partial_{t+1} V_{t+1}} \qquad (1)$$

where V_t = interval velocity of layer t, ∂_t = density of layer t, R_t = reflection strength of interface and t = discrete time index subscript.

The ideal seismogram would be the spike series R_t. The R_t would be zero between the layer boundaries detected by the reflectivity. The earth, however, is a less than ideal transmitting medium with velocities sometimes transisting slowly and acting as a dispersive filter on the higher frequencies. The knife sharp surface impulse is gradually broadened and distorted into a wavelet whose amplitude is only loosely related to the reflection coefficient. Thus the seismogram is a well disguised representation of the reflectivity. It must be interpreted from the wavelets as 'thru a glass darkly.' A second component, an unknown wavelet, is introduced by the earth to the seismogram.

These wavelets, which may be variable in time and space, complicate the interpretation of the data. The wavelets, ω_t, are additively superimposed by the earth to make up the recorded seismogram. The mathematical concept

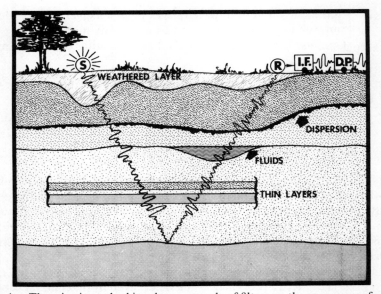

Fig. 1. The seismic method involves a cascade of filters on the source waveform.

is that the wavelets are convolved with the reflection coefficients.

$$f_t = R_t * \omega_t = \int_0^T R_t \omega_{t-\tau} \, dt \qquad (2)$$

where f_t = seismic trace, ω_t = wavelet, R_t = reflections, T = the total time duration of the signal and τ = the time shift.

As the wavelets are overlapped and summed, interference is created between wavelets. Sometimes the interference is constructive, building up spurious amplitudes, and at other times destructive, attenuating reflection responses. The wavelets are seldom simple and the composited seismogram requires skilled interpretation to identify geologic layers even in the simplest lithology.

Some of the factors affecting the shape of the seismic waveform are illustrated in Fig. 1. The source may be impulsive as with dynamite, controlled as with 'Vibroseis' or variant as with weight drop. The earth acts as a cascading filter suite to the source wavelet. The weathered layer, fluids, thin layers, transition zones, erosion, and other earth components tend to attenuate the higher frequencies. Multiples, ghosts, reverberants and other spurious energy complicate the wavefront. The type of geophones used to record the reflected waves affects the shape. Instrumental field recorder

PULSE SHAPING METHODS 243

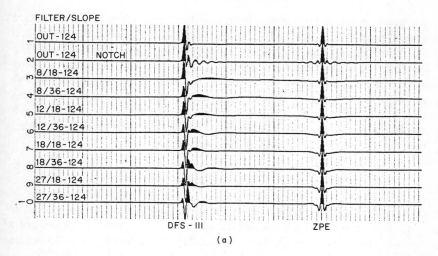

(a)

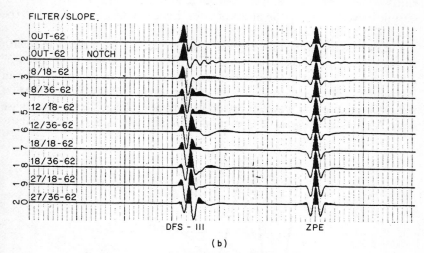

(b)

FIG. 2. DFS-111 filter responses.

filters can vary the phase or shape of the waveform in unique ways. Certain elements of data processing such as geometrical corrections are sources of distortion. The recorded seismic trace may bear small resemblance to the reflection coefficients, A single element of the waveshaping described, such as the instrumental filter, can obscure even the simplest earth layering.

Figure 2 is a suite of DFS-111 instrumental filters. A single reflection coefficient could be represented by any of these wavelets some of which have several peaks. The amplitude of the wavelets would be partially controlled by the reflection coefficients and also by many other mechanisms. The wavelets are up to 100 ms in length so that the response to seismic layering would be a superposition of wavelets. The reflection coefficients and the recorded wavelets are almost inextricably related. For the reflection coefficients the formula of eqn. (1) yields a distinct series of values. No such equation exists for defining the wavelet. Indeed a precise definition of what is meant by a wavelet is not readily proposed.

WAVELETS

A wavelet is a signal whose amplitude is negligibly small except in a finite part of time. A wavelet has a definite epoch or centre while a segment of a seismic trace does not. Robinson and Trietel[1] defined a wavelet as an entity while a trace segment is a part of a larger continuing entity. Typical seismic wavelets are of the order of a few hundred ms as compared to the usual 4–6 s of a seismic trace.

Even the arrival time of a single discrete reflection coefficient varies with the wavelet shape. If the wavelet is symmetrical, the epoch is time zero. For a 90° wavelet it is at the centre but is a zero rather than an epoch. For an instrumental filter or more generally a minimum phase wavelet, it is the first rise. Minimum phase wavelets are difficult to describe as they rotate phase with bandwidth. Figure 3 is a minimum phase wavelet computed by

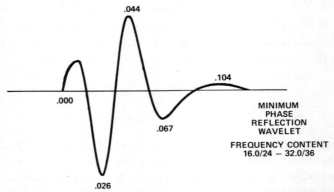

FIG. 3. Minimum phase reflection wavelet; frequency content starting at 16 Hz, damps the spectrum at 24 dB/octave—starting at 32 Hz, decays the spectrum towards Nyquist at 36 dB/octave. From Shugart.[2]

Shugart.[2] Interpretation of seismic data relative to line or well ties are complicated by such characteristics.

The basic concept described is that the seismic trace could be decomposed into a wavelet and a set of spikes closely related to the reflection coefficients. This simplified definition obviously does not include in either part some of the known components of the seismic trace. Multiples of all orders are believed to exist and well log synthetics include such events. Indeed it is even suspected that the short peg leg multiples are what makes the seismic method work. For the seismic trace, then, we can define the following elements:

(1) Wavelet. A signal with a definite epoch but of unknown duration and unknown shape. Moreover this wavelet may change any of its characteristics with time and space.
(2) Reflection coefficients. The ratio of acoustic impedance at layer boundaries described by eqn. (1). These spikes define the primary event boundaries.
(3) Multiples of various types. These include ghosts, reverberation, peg leg and longer multiples.
(4) Noise. Both additive random noise and organised noise, perhaps including diffractions and other structurally generated distortions.

Algorithms for separating these components can make assumptions about these elements and combine them in different ways. The reflection coefficients and the multiples can be combined to create the reflectivity function. Alternatively, some of the types of multiples can be included in the reflectivity and others as part of the wavelet. Assumptions can be made about the wavelet shape and duration, the attenuation of multiples of different periods, and the effect of processing in removing both types of noise.

Clearly the result of such algorithms is to be interpreted in the light of the assumptions made. Decomposition of the seismic trace is then not necessarily unique; it could be decomposed in various ways. This brings out the crux of the interpretation. Is the variation in the seismic response a change in reflection coefficients, wavelet, reflectivity, or noise? With this in mind, a review of wavelet shaping techniques and their development can be useful in evaluating the results from an interpretive viewpoint.

One of the earliest and best known formulations of the seismogram in terms of wavelets was by Ricker.[3] Ricker conducted carefully controlled borehole experiments to record the seismic wavefront. He identified the controlling factors on wavelet shape as the distance travelled and velocity.

Figure 4 shows some of the computed wavelets as a function of travel-time. The energy source was presumed to generate a sharp knife-like pulse changing shape as it proceeded through the earth. His theoretically computed wavelets compared very well with those observed in the borehole.

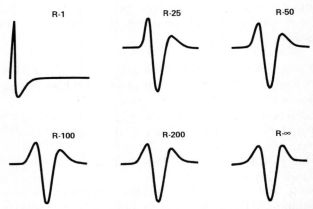

FIG. 4. Ricker wavelets as a function of travel-time. From Ricker.[3]

The theory was the subject of hot debate and widespread interest at the time. Possibly the important contribution of Ricker was the recognition that the seismogram could be described in terms of wavelets superimposed by convolution with the reflectivity function. This concept was useful for the marriage of theoretical signal processing developments and geophysical data processing.

At almost the same time that Ricker was publishing his seismic wavelets, Wiener[4] was publishing his work on linear prediction. Much of his theory was devoted to a wavelet formulation, and the basic concept was that of linear prediction. Using the past history of a signal, a prediction could be made about its future. Unpredictable elements would give rise to a large prediction error and serve to separate correlative and uncorrelative signal components. Seismic responses of the earth can be envisioned in this framework. If the layering is completely predictable, seismic exploration is redundant. But, 'age cannot wither nor custom stale her infinite variety' is a more applicable description of the earth.

In 1953 Robinson[5] made a model of the seismic trace which added complexity to the wavelet. The work of Ricker and Wiener was beautifully combined to attempt the decomposition of the trace into the wavelet and reflection coefficients. To do this the trace was considered to be composed of predictable and unpredictable parts. The reflection coefficients are the unpredictable part (along with additive noise) and the remaining signal the predictable parts. Ghosts, reverberants, and multiples, are predictable and are included as part of the wavelet. Very long multiples, 0·8 s or greater, are just ignored and cannot conveniently be included with the wavelet. Thus the wavelet is not the simple one described by Ricker in this context. The wavelets can be long as contrasted to a reflector sequence, with redundant energy further obscuring the reflectivity function.

Some components of the wavelet are known. The instrumental filter, the published signatures of various marine sources, and 'Vibroseis' recording parameters furnish information about the wavelet. These can be incorporated into filter design and used to reshape the wavelet. Otherwise, the expected wavelet shape is a matter of conjecture. The effort to capture and utilise the wavelet has come about since 1975 with the technological advances in computers and signal processing. The theory of wavelet shaping, however, was developed during World War II by Norbert Wiener. The Wiener shaping filter was described by Wiener[4] and again by Lee.[5] This algorithm is basic throughout the following discussion. It is worthwhile to digress from the wavelet–reflection coefficient theme to review the Wiener–Hopf filter, more commonly called the Wiener shaping filter.

WIENER SHAPING FILTER

The background for the development of the Wiener shaping filter is that of linear prediction. The algorithm itself, however, is just the optimal filter for reshaping any wavelet to a more desirable shape. Given an input wavelet, $i(t)$, and a desired wavelet, $d(t)$, the filter which will transform $i(t)$ to $d(t)$ is

$$h(t) = \frac{\varphi_{i,d}(t)}{\varphi_{i,i}(t)} \tag{3}$$

where $\varphi_{i,i}(t)$ = autocorrelation of the input wavelet $i(t)$ and $\varphi_{i,d}(t)$ = crosscorrelation of the input wavelet, $i(t)$, and the desired wavelet, $d(t)$, so that

$$i(t)*h(t) = d(t)$$

Figure 5 by Mereau[7] illustrates the shaping of a square wave into a semi-circle. With noise-free data, whose spectrum has no zero values, perfect performance can be achieved. As stated, the Wiener shaping filter is deterministic in the sense that the input and output are specified. Generally, for most signal processing applications, the input signal is the transmitted waveform and may be contaminated and not precisely known. Statistical concepts, however, can be brought to bear on the problem as is the case for

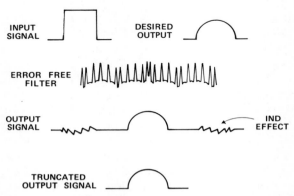

FIG. 5. The transformation of a square into a semi-circle. From Mereau.[7]

seismic data where Robinson[5] explored the convergence of Ricker's wavelet theories and Wiener's statistical approach. In 1957 he proposed an adaptation of linear prediction theory to seismic data.[8] Considered in the context of the shaping filter, the solution becomes a special case with some simplifying assumptions.

The 'input' wavelet was not known for the seismic trace. However, the assumption that the reflectivity function was random paved the way to estimating the autocorrelation of the wavelet from the seismic data. The autocorrelation function measures how well a signal correlates with itself as it is passed by itself in time.

$$\varphi_f(\tau) = \int_0^T f(t)f(t + \tau)\,dt \qquad (4)$$

A completely random signal correlates only at $\tau = 0$ and is uncorrelated thereafter. The trace model is a convolutional equation,

$$f(t) = R(t)*\omega(t) + n(t) \qquad (5)$$

where $R(t)$ = random reflection coefficients, $\omega_{(t)}$ = wavelet and $n(t)$ = additive noise.

The autocorrelation of a segment of $f(t)$ then would collapse $R(t)$ to delay zero and the rest of the $\varphi f(\tau)$ would contain only elements of the wavelet $\omega(t)$. Thus, except for the zero lag, the autocorrelation of a segment of $f(t)$ is the required wavelet autocorrelation of the Wiener shaping filter.

The autocorrelation, $\varphi_{i,i}(t)$ of the Wiener filter is therefore available. The numerator, $\varphi_{i,\,d}(t)$, however, requires that the input wavelet be known. The

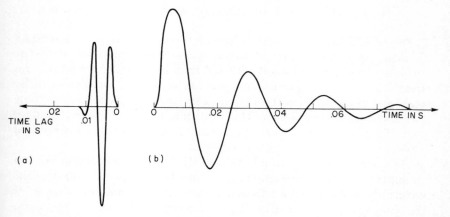

FIG. 6. (a) Inverse wavelet and (b) seismic wavelet. From Robinson.[8]

wavelet being unknown, an assumption of minimum phase was made. Alternatively, the wavelet of Ricker could be assumed as the input wavelet.

With these assumptions the inverse can be computed and such an exercise by Robinson[8] is shown in Fig. 6. Thus the adaptation of the Wiener shaping filter for seismic data involved the following specific assumptions.

(1) The trace is modelled by eqn. (5). The reflectivity series, $R(t)$, is defined to be a random or unpredictable series. High amplitude cyclic bedding can degrade the filter. Fortunately cyclic bedding of the required period and regularity is unusual in nature. Sand-shale and limestone-shale alternations a few inches thick are common, but these are of such high frequency as to be unimportant in practice. Cyclothems in coal measures are from 100 ft thick upwards, and show a rather exact repetition of a stratigraphic sequence, but thicknesses are too variable to meet the requirements.

(2) Since $R(t)$ is uncorrelated, the autocorrelation function of the wavelet, $\omega(t)$, can be estimated by the autocorrelation of a segment of the trace.

(3) The wavelet is assumed to be minimum phase and the desired result a spike. Their crosscorrelation is a scalar vector of the form $\{\alpha, 0, 0, \ldots\}$. This wavelet includes all correlative events such as ghosts, reverberants and multiples within its length. Longer multiples are hopefully attenuated with data processing.

(4) No assumption or provision is made for additive noise except that, being random, its autocorrelation is confined to the first lag.

This set of assumptions formed the basis for waveshaping on seismic data for many years. A minimum phase filter is causal and decays with time. The resulting shape is 'front loaded' with high frequencies leading the waveform and the lower ones in the tail. Computation of the filter was also for many years a challenge to the early computing systems. It came to be commonly called spiking deconvolution. This algorithm proved to be excellent for controlling the frequency variation. The phase and shape of the output wavelet is highly questionable as optimal.

The inversion of the wavelet by the deconvolution filter of Robinson is an attempt to estimate the reflection coefficients. The objective is to reshape the wavelet to a spike. In a noise-free environment this can be done. Figure 7 is the spiking deconvolution of Fig. 3. Unfortunately, the natural earth bandlimiting and noise prevent a complete solution, leaving the reflection

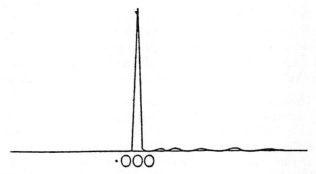

FIG. 7. Spiking deconvolution of pulse in Fig. 3. Unit prediction distance filter length 0·130 s. No prewhitening; no band-limiting. Input Fig. 1 or Fig. 2. From Shugart.[2]

coefficients convolved with an improved wavelet but of questionable polarity and optimality of shape for the usable bandlimits. Neither the reflection coefficients nor the wavelet are successfully separated in practice. The general approach is estimation of the statistics of a short wavelet and inverting it from the data via its autocorrelation.

Spiking deconvolution became the accepted name rather than decomposition as originally given by Robinson. This algorithm has been applied to almost all seismic data since 1968 in some form and, regardless of its deficiencies, it has proved useful as a waveshaper. It stands as a hallmark of geophysical technology. Details of the computational aspects can be found in Appendix 1.

The spiking deconvolution process proved very useful in removing reverberations from marine data as well as collapsing the wavelet. There were problems, however, as it became increasingly evident that the minimum phase assumption was questionable. Certainly for some recorded marine signatures the pulse was seen to be mixed phase. Moreover the random earth assumption sometimes failed as highly reflective, equally spaced layers could dominate the autocorrelation. More recently Berkout[9] has noted that the complete inversion of the spiking deconvolution could generate errors in the phase curve in the presence of noise. Noise being time and space variant, spurious anomalies could be created to confound the interpreter. Robinson[8] described an approach where a 'predictive distance' could be added to the algorithm which would preserve the wavelet but remove reverberants.

PREDICTIVE ERROR FILTERS

Peacock and Trietel[10] received an award for best paper from the SEG for their treatment of predictive error filtering. The basic concept was to create a filter where the first point was unity, a sequence of zeros followed for the prediction gap, and the remainder of the filter performed the usual waveshaping. In this way the basic wavelet could be preserved, and additional control was created for the spectrum whitening, and reverberatory wavelet tails removed. Only a portion of the wavelet is reshaped. If the prediction is unity then the filter is identical to spiking deconvolution.

Using a prediction distance which includes the basic wavelet but not the ghosts and reverberants allows a direct implementation of the Wiener shaping filter. Using the matrix notation of spiking deconvolution gives

$$\begin{bmatrix} \varphi_\alpha & \varphi_{\alpha+1} & \cdots & & \varphi_{\alpha+n} \\ & \varphi_\alpha & & & \\ \varphi_{\alpha+1} & & \ddots & & \\ \vdots & & & \varphi_\alpha & \\ & & & & \ddots & \vdots \\ \varphi_{\alpha+n} & & & & \varphi_\alpha \end{bmatrix} \begin{bmatrix} h_\alpha \\ h_{\alpha+1} \\ \vdots \\ h_{\alpha+n} \end{bmatrix} = \begin{bmatrix} g_\alpha \\ g_{\alpha+1} \\ \vdots \\ g_{\alpha+n} \end{bmatrix}$$

The $\mathbf{G}_t$ is the crosscorrelation of the input and desired waveshape. As desired output is a time-advanced version of the input trace, $\mathbf{X}_t$,

$$\mathbf{G}_t = \sum_t X(t+\alpha)X(t-k) = \varphi(\alpha+k) \qquad (6)$$

which is just the autocorrelation for lags greater than α, the prediction distance. The vector on the right hand side of eqn. (6) becomes $[\varphi_k, \varphi_{k+1}, \ldots]$. The filter can, just as in spiking deconvolution, be computed completely from the autocorrelation and the Levinson recursion can be used to solve the matrix.

Before application the first lag is set equal to unity, zeros are inserted for the prediction distance, and the $h(t)$ negated. The filter then is the vector $[1\cdot0, 0, 0, \ldots -f_k, -f_{k+1}, \ldots, -f_n]$. It is obvious that the wavelet will be copied for k lags and deconvolved thereafter. Figure 8 shows an example of

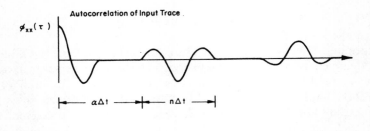

FIG. 8. Typical input and output autocorrelation. From Peacock and Trietel.[10]

the performance of the filter as measured by the autocorrelation. The input autocorrelation of the trace is compared to the output after application of the filter. The prediction distance preserves the function and the wavelet tail is removed thereafter.

This filter remains a mainstay of geophysical data processing. For bright spot or direct detection exploration it is much to be preferred to spiking deconvolution as it does not affect polarity. With many energy sources, such as 'Vibroseis', the minimum phase assumption is in doubt and prediction avoids distortion of the basic wavelet. This method introduces no new phase distortion but does not standardise the wavelet shape. The shape is therefore not optimal.

Both types of deconvolution avoid the basic problem rather than solve it. Spiking deconvolution assumes the wavelet is minimum phase. Prediction error filtering just passes the wavelet shape intact for the prediction distance. Neither fully implements the Wiener shaping filter because of lack of information about the wavelet shape. An increasing interest in detailed stratigraphic interpretation generated new approaches around 1975.

Marine recording created several new sources such as air-gun, 'Vaporchoc', etc. In many cases the signature of the source was known. This analytic portion of the seismic trace can be completely optimised by the Wiener shaping filter. The availability of Fast Fourier computer hardware motivates a return to the frequency format originally used by Wiener and Lee.

DETERMINISTIC DECONVOLUTION

Let $s(t)$ denote the signature of the source. The shaping filter then is

$$H(\omega) = \frac{\Phi_{sd}^{(\omega)}}{\Phi_{ss}^{(\omega)+c}} \quad (7)$$

where c = arbitrary constant, $\Phi_{ss}^{(\omega)}$ = power spectrum of $s(t)$ and $\Phi_{sd}^{(\omega)}$ = crosspower of $s(t)$ and $d(t)$ in Fourier transform. The desired result is a spike collapsing the numerator to

$$H(\omega) = \frac{S*(\omega)}{\Phi_{ss}^{(\omega)+c}} \quad (8)$$

where $S*(\omega)$ = conjugate of $s(\omega)$. The calculation then is to crosscorrelate with the signature and divide by its power. This operation removes source created reverberation and phase distortion from the source waveform.

Another case where the shaping filter can be exactly applied is the removal of instrumental filters applied by the seismic recording equipment. These instrumental filters are typically of a Butterworth design and have a minimum phase component. Figure 2 from Neale[11] shows some of the filter responses measured from a field recorder. These filters have a complex response to a simple spike and degrade the interpretability of the data. The instrumental filters can be removed from the data by eqn. (8) which reshapes the phase curve to zero.

The instrumental filters can be obtained by putting a spike through the field system and recording the response. In practice, the majority of the field instrumental filters can be easily synthesised by a Butterworth filter design. Often equipment manuals will include a phase and amplitude spectrum plot of the filter.

Any known phase component can be removed deterministically. For instance, many 'Vibroseis' recording systems feature a phase compensation network to monitor the output vibration. These networks are usually 90° out of phase and a compensating filter can be easily computed to remove the phase shift. Where the wavelet component is known, it can be completely corrected for phase distortion. However, the denominator of eqn. (7), the power spectrum, may contain spectral zeros. In this case the algorithm must, just as in standard deconvolution, be protected against zero division.

WAVELET PROCESSING

In 1975 the increased amount of detailed seismic stratigraphy and the increased borehole data motivated what is now known as wavelet processing. Schoenberger and Levin[12] demonstrated that the most desirable wavelet shape was the zero phase waveform. This waveform, relative to any other shape, has the best peak–side lobe ratio. Arrival time is at the more easily detected peak rather than at the onset as with minimum phase waveforms. If the waveform is detected and reshaped to a known form one of the interpretational variables will be removed from the data.

Barry and Shugart[13] discussed the capture of marine source signatures from deep water. If the water bottom was deep enough so that its first multiple was not interfering, the first breaks would be an estimation of the source signature. This procedure required careful control and had limited applicability but was demonstrated to improve the data. Signature capture was still largely deterministic and accounted for only one distortion in the waveform.

Lindsey et al.[14] proposed extracting the wavelet by integrating the response to a known layer. White et al.[15] used minimum phase versions of averaged autocorrelations. Oppenheim and Schafter[16] of the Massachusetts Institute of Technology motivated efforts to extract the wavelet with a homomorphic approach. Ulrych,[17] Stone,[18] and others attempted implementation of the method on seismic data. The target of all these efforts was to fully implement the Weiner shaping filter to achieve a zero phase wavelet with a level spectrum over its bandwidth.

A method of a pure statistical approach was proposed by Stone.[19] The approach is called structural deconvolution and is applicable to any seismic data regardless of source, recording location, type of coverage, or data quality. Deconvolution attempts to directly estimate a wavelet and compute a deconvolving filter without phase information. An alternative approach is to directly estimate the reflection coefficients and deconvolve them from the trace leaving the wavelet. With this method the phase of the wavelet is preserved removing the need for a phase assumption. Successful capture of the wavelet brings seismic waveshaping full circle to direct implementation of the Wiener shaping filter. Moreover, the reflection coefficient estimates are available for processing and interpretation.

STRUCTURAL DECONVOLUTION

To accomplish estimating the reflection coefficients a more flexible trace model is assumed

$$f(t) = R(t)*[\omega(t)*h(t)] + n(t) \qquad (9)$$

where $R(t) \sim$ reflection coefficients and long period multiples, $\omega(t) =$ source wavelet including ghosts and reverberants, $h(t) =$ time variant earth filter and $n(t) =$ additive random noise. The reflectivity, $R(t)$, remains the unpredictable uncorrelative part. The wavelet, $\omega(t)$, is now allowed to be continuously time variant as modified by the earth filter, $h(t)$. The additive noise term is included in the model and will be recognised in the estimation process.

The structural deconvolution method for capturing the wavelet is to estimate the reflectivity and invert it from the trace. A familiar exercise is to crosscorrelate well log synthetics with corresponding seismic traces to extract the wavelet. The estimation of reflectivity for the same purpose gives a general statistical solution where well logs are not available. Moreover, a unique wavelet can be derived for each trace to accommodate lateral variance.

Estimation of the reflectivity is the difficult part of the approach. It is a subject of sparse literature and of recent technology. To implement the model of eqn. (9) some characteristics of the estimation are defined. The time–space variance allowed implies an algorithm which is rapidly adaptive in both dimensions. The algorithm should be based on linear prediction to extract the uncorrelated reflectivity. The additive random noise, $n(t)$, is expected to be unpredictable but also uncorrelated trace to trace. A multichannel procedure could separate the noise from the reflectivity. An algorithm with these characteristics is the predictive adaptive multichannel maximum entropy algorithm. A popular name for the basic algorithm is MESA (maximum entropy spectral analysis). John Burg[20] developed this approach as a high resolution spectral analysis technique.

The important differences between MESA in an adaptive form and traditional deconvolution as previously defined are:

(1) Both a forward and backward prediction error filter are computed. These correspond to the upgoing and downgoing wave fronts;
(2) The autocorrelation function is not used so that the possibility of creating spurious amplitude spectrum values from its truncation are avoided. The data samples are used directly in prediction;
(3) The adaptive mode relaxes the stationary assumption required by deconvolution, adapting rapidly to change with time;
(4) More information is generated. A forward and backward prediction error series is generated along with a stable reflection coefficient vector.
(5) The output is completely broad-band, if desired, so that a spike series is generated rather than a partially compressed wavelet. This is because the randomly arriving coefficients must have non-zero spectrum.

The mathematics of MESA have been widely documented. The most complete treatment is by Burg.[20] Other helpful sources are Claerbout,[21] Anderson,[22] and Barnard.[23] The basic concept is still that of Wiener.[4] A prediction is made based on the previous input trace values and a prediction error generated. Let

$$\mathbf{X}_t = \{X_1, X_2, \ldots X_T\}$$

be the input trace. The prediction error is

$$\mathbf{E}_t = X_t - \hat{X}_t \qquad (10)$$

where $\hat{X}_t$ is the predicted value. This error is minimised differently from

Wiener by using a forward error E_t and a backward error B_t. The minimising value is for an n length filter. For the ith iteration the equation is

$$C_{i+1} = \frac{-2 \cdot 0 \sum E_t B_t}{\sum (E_t^2 + B_t^2)} \qquad t = 1, 2 \ldots, n \qquad (11)$$

This is recognisable as the familiar coherence function. Claerbout[21] presents arguments relating C_t to the reflection coefficients. The coherence between upgoing and downgoing waves is a wave theory migration definition of a seismic reflection. The C_t is always less than unity and is always stable.

The forward and backward error series are recursively updated using the C_t as

$$E_{i+1} \leftarrow E_i + C_{i+1} B_i$$
$$B_{i+1} \leftarrow B_i + C_{i+1} E_i \qquad i = 1, 2, \ldots, n \qquad (12)$$

and the prediction error filter by

$$P_{i+1} \leftarrow P_i + C_{i+1} P_i^*$$

with

$$P_i \leftarrow 0 \cdot 0 \qquad \text{and} \qquad P_0 = 1 \cdot 0 \qquad (13)$$

The adaptive mode uses the above algorithms to initialise a filter of length n. As each new trace value arrives it is used to modify the above vectors. Barnard,[23] Griffiths et al.,[24] and Burg,[20] give good explanations of an adaptive algorithm. Riley and Burg[25] present some results on real data based on these equations.

To implement the structural deconvolution approach an additional assumption is made. The multichannel predictive adaptive Burg algorithm is assumed to detect enough of the reflection coefficients to serve as a wavelet estimation tool. This assumption is well tested both synthetically and in practice on seismic data. The structural deconvolution approach to extracting the wavelet is as follows.

I. Estimate the reflection coefficients with the MESA method in an adaptive and multichannel mode,

$$\hat{R}(t) = M[f(t)]$$

where $\hat{R}(t)$ = estimated reflectivity.

II. Crosscorrelate the estimated reflection coefficients with the seismic trace to extract the wavelet,

$$\omega(t) = f(t) * \hat{R}(t) \qquad (14)$$

The crosscorrelation will be as long as the trace, an unreasonable wavelet length. Conventional tapering functions to limit the wavelet length are applied so that the standard deconvolution filter lengths are possible. The tapering also eliminates background correlation and residual structure.

If the seismic section is long in time and significant time variance is suspected, gates on $\hat{R}(t)$ and $f(t)$ can be imposed just as in conventional deconvolution. The well established power of the MESA approach on short time gates makes time variant wavelet processing much more feasible.

Just as in 'Vibroseis' crosscorrelation, scaling of the estimates does not affect the wavelet shape. Errors on the very small reflection coefficients tend to average out. The multichannel reflection coefficient algorithm suppresses noise in $\hat{R}(t)$ and in the estimated $\omega(t)$. Only highly correlated high amplitude structure can induce significant error. Introduction of the autocorrelation as a denominator could give protection but is not generally needed.

III. Use the estimated wavelet to compute a shaping filter by the Wiener–Hopf definition,

$$h(t) = \frac{\omega*(t)}{\varphi\omega(t)}, \qquad \varphi_\omega(t) = \text{autocorrelation of } \omega(t) \qquad (15)$$

and apply $h(t)$ to the trace

$$f_z(t) = f(t)*h(t) \qquad (16)$$

so that $f_z(t)$ is the seismic trace with a symmetrical wavelet. The crosscorrelation of the wavelet with itself gives a zero phase pulse. The denominator levels the spectrum in the non-zero portion.

Figure 9 depicts the structural deconvolution process. The input trace is followed by the reflection coefficients as measured by MESA. These may be compared to the actual coefficients used to generate the input trace. Most of the discrepancies are scalar or on the smaller amplitudes. The crosscorrelation of the estimates with the input as described in eqn. (14) produces the estimated wavelet. It may be compared to the actual wavelet. The errors in estimation do not appreciably affect the wavelet estimation.

For seismic data the process holds the capability of improving the interpretation. The wavelets are estimated time and space variantly and applied by eqn. (15). Possible interpretive benefits which could result are:

(1) Better resolution of closely spaced events such as pinchouts and thin sands due to wavelet optimisation;

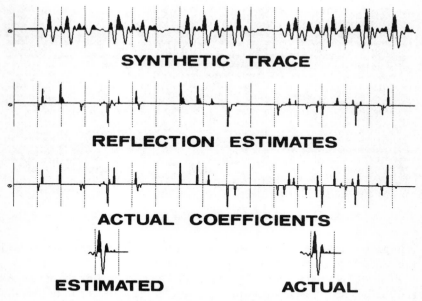

FIG. 9. Pulse estimation by structural deconvolution.

(2) Elimination of time or spatial variance in the wavelet which might be interpreted as subtle stratigraphic changes;
(3) Easier correlation with well log synthetics due to the easily generated convolving wavelet;
(4) Improvement in line times where differing instrumental filters, geophones, processing, etc. are involved.

Some of these benefits are illustrated by data from Michigan, USA. The targets in the area are small, often dim, reef structures in the Niagaran group. Figure 10 has an arrow pointing toward a reef which is, in this case, clearly visible. The overlying positive (black) event is the A-2 carbonate and the underlying is the Niagara. Note the changing character from one side of the reef to the other, as well as to the left end. Are these stratigraphic or due to changes in the rather thick weathered layer common to the area? Figure 11 shows the reflection coefficient estimates. The estimates map the A-2 and Niagara as the same horizontally across the section, except at the gas bearing reef.

Figure 12 is the wavelet processed section. Like the reflection estimates the two events are clearly seen to be laterally continuous. The reef definition

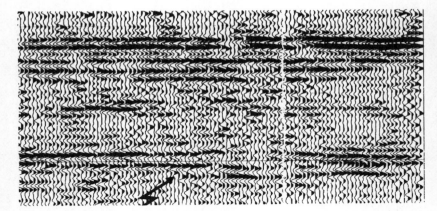

FIG. 10. Standard processing stacked section and input to wavelet processing.

is improved and the low amplitude effect of the reef is more visible. Arrival times are at peaks, easing the layer identification.

Figure 13 is a section where a thick chalk layer is noted by the arrow. As the chalk is nearly 80 m thick, the signature of the wavelet can be visually verified. The expanded scale shows wavelets extracted by the method. The response to the top of the chalk should have been a high amplitude positive. Instead, a mixed-phase negative response is seen highly similar to the extracted wavelets. Note also the lateral character variance. The output section (Fig. 14) clearly demarcates the chalk boundary and removes most

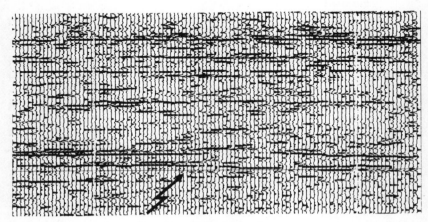

FIG. 11. Reflection estimates.

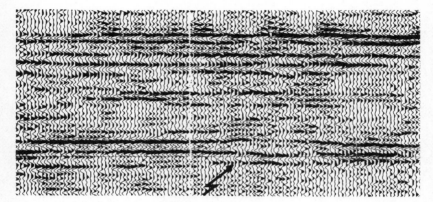

FIG. 12. Wavelet processed section.

of the spurious lateral change. The expanded scale insert reflects the improvement in waveshape enhancing the interpretation of the thin target sands deeper in the section. Events which are not symmetrical represent more than one reflection coefficient. The detection of the appearance or disappearance of thin sands is made more feasible by the symmetrical waveshape adding this dimension to the display.

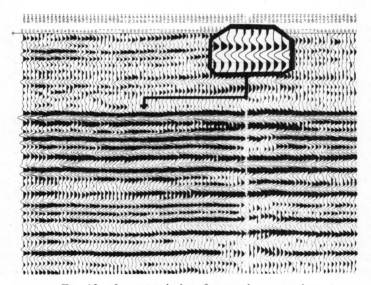

FIG. 13. Input stack data for wavelet processing.

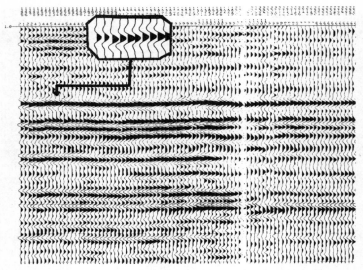

FIG. 14. Wavelet processed result for Fig. 13.

The comparison of the data and the well log synthetic illustrates the utility of establishing zero phase (Fig. 15). The synthetic was filtered with a simple bandpass to achieve the character match. An interesting by-product of the method is the estimation of reflection coefficients; the assumption is that the estimates are adequate for wavelet extraction. The surprising quality of the estimates motivates a consideration of their utility in interpretation. The original requirement is, in some cases, more modest than the results indicate necessary.

ESTIMATION OF REFLECTION COEFFICIENTS

The ideal waveshape would be a spike. Even if the wavelets were collapsed to a spike many more conditions must be met before the reflection coefficients would be realised from a seismic trace. Characteristics of the seismic reflection method such as geometrical corrections, earth filtering, redundancy, and dipping layers, along with other factors, obscure the reflectivity. These differences in acquisition cause the borehole and seismic reflectivity to differ. Moreover, boreholes are usually over some anomalous part of the data. The estimation of reflectivity must be restrained to that

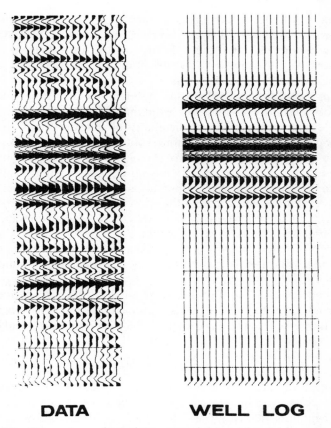

DATA   **WELL LOG**

FIG. 15. Comparison of wavelet processed data and well log synthetic.

measurable by the seismic data. Under conditions of low noise and high upper frequency limits there should be a good deal of coincidence between the well log and seismic data.

The basis of the MESA method previously described is decomposition of the trace into predictable and unpredictable parts. Strongly cyclic sediments will escape the algorithm but perfect predictability is relatively rare. The previous discussion of the algorithm only strove to present the estimates as sufficiently good for wavelet extraction. To be used for interpretive purposes a stronger demand and a more extensive process is needed. One such extension is to monitor the overall coefficient estimate trace gain.

When a well log synthetic is available on the line, the reflection coefficients of well defined layers can be used for scaling. On marine data the reverberants give a strong indication of the water bottom reflection coefficient. Any interpretive knowledge of the area can of course be used to improve the reflection coefficients.

The predictive–adaptive algorithm of Burg[20] has already been described. The most essential ingredient for estimation of the reflection coefficients from a seismic trace is the predictability of the wavelet. A supporting statistic is the validity of the assumption of the primary events being weakly correlated. As the adaptive operator is typically only 10 or so samples, this assumption is not too restrictive. Only closely spaced highly cyclic layers will escape the operator. As the operator is analysing waveforms, the well-known resolution limit for detecting closely spaced events is applicable. Widness[26] and others have concluded that the waveform does not change shape after the events close to within $\frac{1}{8}$ wavelength of the dominant frequency. Clearly this or any algorithm will vary in success as the signal–noise ratio and the correctness of the statistical assumptions, but will be limited by the data parameters.

The concept is straightforward, depending on the convolutional trace model and wavelet processing. It is presumed that wavelet processing has successfully established a basic wavelet which is symmetrical, and the trace is of the form,

$$f_z(t) = R(t)*b(t)$$

where $b(t)$ is the bandlimited zero phase wavelet and $R(t)$ is the reflectivity.

The reflectivity estimates, $\hat{R}(t)$, are a spike series. The wavelet, $b(t)$ is a simple bandpass filter easily designed from amplitude spectra of $f_z(t)$. The only quality control on $b(t)$ being symmetrical is to use the structural deconvolution to extract the wavelet from $f_z(t)$. Suppose the reflectivity estimates, $\hat{R}(t)$ are filtered with $b(t)$, then

$$\hat{f}(t) = \hat{R}(t)*b(t)$$

Thus two versions of the trace are generated, $f_z(t)$ and $\hat{f}_t$. Each is a spike series convolved with the same or at least a very similar wavelet. The wavelet processed and filter estimate traces are

$$f_z(t) = R(t)*b(t)$$

and

$$\hat{f}(t) = \hat{R}(t)*b(t)$$

so that if $f_z(t) = \hat{f}(t)$, then obviously $\hat{R}(t) = R(t)$ and the reflectivity

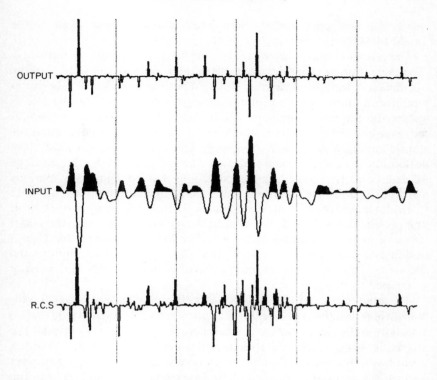

FIG. 16. A set of reflection coefficients, a filtered version for input, and the estimates made.

estimates are correct. Of course it is expected that $\hat{R}(t) \approx R(t)$, i.e. approximately, but not exactly equal. The point is that with $f_z(t)$ and $\hat{f}(t)$ having a common and easily defined wavelet, $\hat{R}(t)$ can be visually or analytically evaluated by comparing $f_z(t)$ and $\hat{f}(t)$.

The amplitude envelope of the wavelet processed data can be computed. It is presumed that the bandlimited data has been properly processed with regard to gain. Correction of the gain curve of the estimates to that of the input can improve the utility of the estimates.

A suitable model meeting the statistical requirements is displayed in Fig. 16 as a spike series denoted as reflection coefficients series (R.C.S.). This series synthesises the reflection coefficients. A minimum phase filter was applied to create the trace as input to the estimator. The filter has an upper transition point at 25 Hz and a truncation to zero at 70 Hz, closely simulating field instrumental filters. The trace is the input signal to the modified MESA algorithm.

The estimates may be compared to the input to evaluate performance. The estimation has considerable merit and has detected about 85% of the significant reflection coefficients. It should be noted that Wiener–Levinson deconvolution, or any algorithm involving division, would not be able to extend the frequency content past 70 Hz as the spectrum is zero and cannot be inverted. The estimates in this case are certainly of enough merit for consideration in stratigraphic mapping. This example does however have little additive noise (1%) and the bandwidth is perhaps better than average seismic data. It is nonetheless encouraging that no significant spurious spikes arise and that the major peaks are all accurately extracted.

Evaluation of the estimator performance on real data is subjective even in the presence of well logs. The proposed method was to make the initial estimate of the coefficients by the adaptive MESA as in Fig. 16. Then a residual scaling was to be made from the input data to improve the estimates. The improved estimates could then be filtered and visually compared to the data.

The input record of Fig. 17 is followed by the reflectivity estimates. A well-known limestone layer just below 1 s appears to be crisply defined on the top and bottom by the estimates. A small negative just above it indicates the overlying shale. Using the structural deconvolution method, the input record was processed to have a zero phase wavelet. The wavelet processed record was checked to be zero phase by extracting the wavelet with the original estimates. Thus it is composed of a spike reflectivity convolved with a symmetric pulse. The top of the limestone is an appropriately strong positive response giving additional confidence. While extracting wavelets is challenging, the computation of the trace amplitude spectrum is a standard processing task. Computing the amplitude spectrum of the wavelet processed data provides the information to design a zero phase bandpass filter.

The R.C. estimates are filtered with the zero phase filter and follow the wavelet processed record in Fig. 15. If the estimates are correct and the amplitude spectrum of the filter matched, then the wavelet processed data and the filtered estimates will be identical.

$$f(t) = R(t)*b(t)$$
$$\hat{f}(t) = R(t)*b(t) \tag{17}$$

where $f(t)$ = filtered estimates. Clearly if $\hat{f}(t) = f(t)$ then $R(t) = \hat{R}(t)$. In this case they are very similar and only minor discrepancies can be seen. The

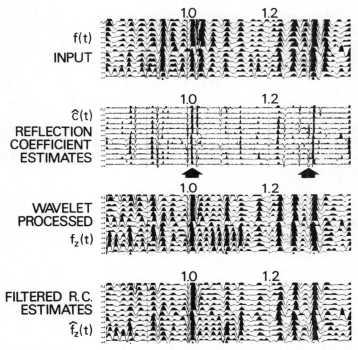

FIG. 17. Comparison of wavelet processed record and filtered reflection coefficient estimation.

conclusion is that the estimates closely approximate the reflectivity. Each layer should, however, be independently evaluated using these displays. If the differences between the two signals are below the visual range, most of the reliable information about the reflection coefficients has been extracted from the seismic data.

APPENDIX 1: THE SPIKING DECONVOLUTION ALGORITHM

The assumptions made about the seismic signal to implement the spiking deconvolution result in the earth model

$$f(t) = R(t)*\omega(t) + n(t) \qquad (A1)$$

where $R(t)$ = random reflectivity, $\omega(t)$ = minimum phase wavelet and $n(t)$ = additive noise. The actual implemented algorithm requires considerable computation and the details were outlined by Robinson[27]. Calculation of the autocorrelation is done over a selected time gate of the trace.

$$\varphi(\tau) = \sum_{T_1}^{T_2} f(t)f(t+\tau) \quad \text{(A2)}$$

An optimal tapering of $\varphi(\tau)$ on its end improves its invertability. Even so, the truncation of $\varphi(\tau)$ to normal filter lengths can generate spectral zeros. Additional insurance for stability is made by adding a small percentage (usually 5% or less) to $\varphi(0)$,

$$\varphi(0) = \varphi(0) + a\varphi(0) \quad \text{(A3)}$$

where a = percentage.

As the result in the frequency spectrum is addition of a constant, the process is called 'adding white noise'. In matrix form the result is an increased diagonal dominance. Diagonal dominance is another criterion for invertability. A matrix is formed as

$$\begin{bmatrix} R_0 & R_1 & \cdots & & R_n \\ R_1 & R_6 & \cdots & & R_{n-1} \\ \vdots & \vdots & R_0 & & \vdots \\ R_n & R_{n-1} & & & R_0 \end{bmatrix} \begin{bmatrix} h_1 \\ h_2 \\ \vdots \\ h_n \end{bmatrix} = \begin{bmatrix} \alpha \\ 0 \\ 0 \\ \vdots \\ 0 \end{bmatrix} \quad \text{(A4)}$$

where R = autocorrelation of the trace, h = spiking filter and α = scalar output of minimum phase crosscorrelation.

Levinson[28] provided a highly efficient recursive solution which is widely used. The filter solution is often called the Wiener–Levinson algorithm. The output filter, $h(t)$ is minimum phase in accordance with the original assumption. Application of $h(t)$ to the trace is by convolution, producing $\hat{f}(t)$, the convolved trace,

$$\hat{f}(t) = f(t) * h(t)$$

Often $\hat{f}(t)$ is very noisy in the higher frequencies so that a low pass filter must be applied. Robinson also noted that an alternative would be to invert only the tail of the autocorrelation.

REFERENCES

1. ROBINSON, E. A. and TREITEL, S., Seismic wave propagation in terms of communication theory, *Geophysics*, **31**, p. 17, 1966.
2. SHUGART, T. R., *Deconvolution; an illustrative review*, presented at 45th Annual International SEG Meeting, Denver, Co., USA. SEG preprint, Box 3098, Tulsa, Okla, USA, 1975.
3. RICKER, N., Wavelet contraction, wavelet expansion and control of seismic resolution, *Geophysics*, **18**, p. 769, 1953.
4. WIENER, N., *The extrapolation, interpolation, and smoothing of stationary time series with engineering applications*, John Wiley and Sons, New York, NY, USA, 1949.
5. WADSWORTH, G. P., ROBINSON, E. A., BRYAN, J. G. and HURLY, P. M. Detection of reflections on seismic records by linear operators, *Geophysics*, **18**, 1953.
6. LEE, Y. W., *Statistical theory of communication*, John Wiley & Sons, New York, NY, USA, 1960.
7. MERFAU, R. F., Exact wave-shaping with a time-domain digital filter of finite length, *Geophysics*, **41**, p. 659, 1976.
8. ROBINSON, E. A., Predictive decomposition of seismic traces, *Geophysics*, **22**, p. 482, 1957.
9. BERKOUT, A. J., Least squares inverse filtering, *Geophysics*, **42**, 1977.
10. PEACOCK, K. L. and TREITEL, S., Predictive deconvolution theory and practice, *Geophysics*, **34**, 1969.
11. NEALE, G. H., *Effects of field recording filters on seismic wavelets*, SEG preprint, Box 3098, Tulsa, Okla, USA, 1977.
12. SCHOENBERGER, M. and LEVIN, F. K., Resolution comparison of minimum phase and zero phase signals, *Geophysics*, **39**, p. 826, 1974.
13. BARRY, K. and SHUGART, R. T., *Zero phase seismic sections*, SEG preprint, Box 3098, Tulsa, Okla., USA, 1975.
14. LINDSEY, P., NIEDELL, N. and HILTERMAN, F., *Interpretive uses of seismic modelling with emphasis on stratigraphy and seismic resolution*, SEG preprint, Box 3098, Tulsa, Okla., USA, 1975.
15. WHITE, R. E., O'BRIEN, P. N. S. and LUCAS, A. L., Estimation of the primary seismic pulse, *Geophys. Prospecting*, **22**, p. 627, 1974.
16. OPPENHEIM, A. V. and SHAFTER, R. W., *Digital signal processing*, Prentice-Hall, Inc., Englewood Cliffs, NJ, USA, 1975.
17. ULRYCH, T. J., Maximum entropy power spectrum of truncated sinusoids, *J. Geophys. Res.*, **77**, No. 8, p. 1369, 1972.
18. STONE, D., *Estimation of reflection coefficients from seismic data*, SEG preprint, Box 3098, Tulsa, Okla., USA, 1977.
19. STONE, D., *Robust wavelet estimation by structural deconvolution*, SEG preprint, Box 3098, Tulsa, Okla., USA, 1976.
20. BURG, J. P., *Maximum entropy spectral analysis*, PhD dissertation, Stanford University, Stanford, Ca, USA, 1976.
21. CLAERBOUT, J. F., *Fundamentals of geophysical data processing*, McGraw-Hill, New York, NY, USA, 1976.
22. ANDERSON, N. O., On calculation of filter coefficients for maximum entropy spectral analysis, *Geophysics*, **39**, p. 69, 1974.

23. BARNARD, T. E. Technical Report No. 1; *The maximum entropy spectrum and the Burg technique*, Texas Instruments, Inc., Dallas, Tx, USA, ALEX(03)-TR-75-01, 1975.
24. GRIFFITHS, L. J., SMOLKA, F. R. and TREMBLY, L. D., Adaptive deconvolution: a new technique for processing time variant seismic data, *Geophysics*, **42**, p. 742, 1977.
25. RILEY, D. and BURG, J. P., *Time and space deconvolution filters*, SEG preprint, Box 3098, Tulsa, Okla., USA, 1972.
26. WIDNESS, M. B., How thin is a thin bed? *Geophysics*, **38**, p. 1176, 1973.
27. ROBINSON, E. A., Multichannel Z-transforms and minimum delay, *Geophysics*, **31**, 1966.
28. LEVINSON, N., Wiener r.m.s. error criterion in filter design and prediction, Appendix B in *Extrapolation, interpolation, and smoothing of stationary time series* (Wiener, N.), Technology Press of MIT, Cambridge, Mass, USA, 1947.

Chapter 8

SEISMIC PROFILING FOR COAL ON LAND

A. ZIOLKOWSKI

National Coal Board, London, UK

SUMMARY

Modern deep underground coalmining technology demands the existence of convenient areas of undisturbed coal for continuity of production. Unexpected geological structural discontinuities can dramatically increase mining costs by halting production locally for months at a time. High resolution seismic surveys can be used to reduce costs by identifying areas of coal with a low intensity of structural disturbance.

Since all coal presently accessible to modern mining technology is shallower than about 1500 m, the conventional oil-scale seismic reflection system must be scaled down to focus on these horizons; higher frequencies must also be used. This inevitably emphasises the distorting influence of the near-surface on the reflection data and special consideration must therefore be devoted to calculation of static corrections and removal of ground roll.

The limits to the resolution of minor faulting are controlled by the magnitude of the static errors rather than by the bandwidth of the data.

INTRODUCTION

It is worthwhile considering why the seismic reflection technique could have any application to problems in coalmining before discussing that application in any detail. The technique has, after all, been developed primarily to reduce the costs of exploration for oil and gas. And since there has never been any problem in finding coal, it is not immediately obvious why seismic reflection could help the coal industry. In fact, the need for it arises not so much from its ability to help locate the reserves, but more from its ability to help reduce the costs of extraction.

In the oil and gas industries the problem is to locate the reserves and to pinpoint the best place to drill in order to extract them. Seismic reflection is used to delineate the geological structure. Whether any of the strata within the structure actually contain oil or gas cannot normally be determined from the seismic data alone. Drilling is required. Only in exceptional cases can the presence of hydrocarbons be inferred from the seismic data— perhaps from an amplitude anomaly associated with a gas–oil contact. But even in such exceptional cases, the proof of the hydrocarbons is in the drilling, and not in the seismic data.

However, once the seismic horizons have been correlated with geological horizons obtained from well data, the seismic data can be used to determine the approximate size of the field and the best places to drill for extraction. As far as the search for oil and gas is concerned, the great power of the seismic reflection technique is that it permits identification of those places where drilling is likely to be successful if hydrocarbons are present, and also to identify those other places where drilling is likely to be quite unsuccessful. Without the use of seismic reflection the probability of any well being successful is greatly reduced, and of course the reason it is so important to increase the probability of the success of any well is that drilling is so costly.

Any exploration programme should be designed to obtain sufficient information about the geological environment at the minimum cost. In general a seismic survey by itself does not provide enough positive information. Usually at least one borehole is also required. However, it is normally far cheaper to obtain the required amount of geological information by using a combination of boreholes plus seismic surveys rather than by using boreholes alone. Essentially, then, in the oil and gas industries, seismic reflection is employed to reduce the cost of exploration.

Once the exploration has been completed, a well can be drilled in an appropriate place and extraction is reasonably straightforward, conceptually. The oil or gas, being fluid, will flow up the well to the surface under the differential hydrostatic pressure within the earth. If this differential pressure is not great enough, a new hole can be drilled in a more appropriate place, or the process may be aided by pumping.

Coal differs from oil and gas in two important respects; first, there is really no difficulty in finding it and, secondly, being solid, it has to be mined. In certain geological environments the costs of mining can be extremely sensitive to the geological structure. It is in these environments that seismic reflection surveys may be used to determine the nature and location of major structural features in the coalfield which would, if discovered only in the course of extracting the coal, cause disruptions in the continuity of mining and result in increased costs. In order to appreciate what these costs might be, relative to the costs of seismic surveying, it is necessary to review briefly the nature of modern coalmining methods and the extent to which these are constrained by the geological environment.

If the coal is very shallow or outcrops at the surface, it may be extracted by opencast or strip mining, in which the shallow overburden is stripped off to leave the coal exposed and ready for extraction by further stripping. Mining by opencast methods accounts for about 10% of current annual extraction in the UK. As the thickness of the overburden increases, the cost of stripping it off to expose the coal also increases. When the overburden becomes too thick for economic opencast mining, it is necessary to use deep mining methods.

If the coal is not too deep, it may be mined by bord and pillar methods, in which some fraction of the coal is mined out at depth, while the remaining fraction is left in the ground in the form of pillars to support the overburden. The deeper the coal is, the larger the fraction which must be left in the ground for this support, and the smaller the fraction which is available for extraction. Therefore, for a given volume of coal output from the mine, the deeper the coal, the further the workings must extend from the point of access to it. Now the cost of shafts, tunnel supports, conveyor systems, ventilation, etc. increases both with the depth of the coal and with the distance of the workings from the shafts. Therefore depth is a financial constraint on bord and pillar mining, because the percentage extraction decreases with depth while the capital costs increase with depth, and a depth is reached at which mining in this way is uneconomic.

The method of mining which permits a much higher percentage extraction at depths which are uneconomic to bord and pillar methods is known as longwall mining. About 80% of British coal is won from highly mechanised longwall faces. The main features of a longwall coal face are illustrated in Fig. 1. A rectangular panel of coal is extracted by moving a tunnel sideways between two parallel roads. These roads are linked to the main underground network, and thence to the surface. The tunnel at the coal face is formed between the coal on one side and the roof supports on the other, as shown in Fig. 2. The supports have cantilevered roof shields which protect the men, coal cutting machinery, and face conveyor system. As the coal is cut away from the face in slices, the coal cutting machinery, face conveyor system and roof supports advance, allowing the overburden to subside behind and form the goaf. (See Fig. 3).

This is a very inflexible system. All the coal face machinery, which can cost up to £2 million and can weigh well over 1000 tonnes, is trapped

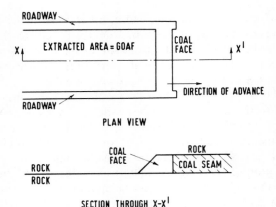

FIG. 1. Sketch to show the main features of a longwall coal face.

between the overburden, the goaf and the face. As long as the seam is continuous in the direction of advance this is no disadvantage; indeed the system is designed exactly to meet this situation. But as soon as the seam continuity is broken (by faulting, by buried sandstone channels, or by seam splitting, for example) all the face machinery is stuck in a narrow tunnel where it cannot be easily manoeuvred and where it can no longer pay for itself by mining coal. It is clear that for this method of mining to work, it is necessary

(1) for the geological disturbances to be so arranged that it is possible to fit the rectangular panels between them, and
(2) to know how the geological disturbances are arranged so that the rectangular panels can be fitted between them.

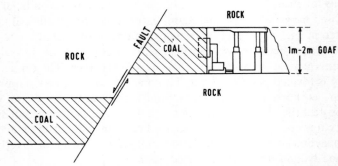

FIG. 2. Sketch to show in profile a coal face advancing towards a normal fault.

Fig. 3. A view down a longwall face; roof supports on the right, the coal face on the left, and the coal cutting machinery straight ahead and moving away from the viewer.

Typically a colliery producing 1 million tonnes of coal per annum will have as few as three or four working faces. To pay off the cost of setting up a coal face a minimum panel run is required. However, in general, it is not normally known whether the ground ahead of any face is sufficiently undisturbed to allow the face to advance the required distance. Normally any fault with a throw of seam thickness or greater will stop a face, while faults with throws less than this will slow down production. Figures 2 and 4 show a possible face-stopping fault ahead of a working face. Once the fault

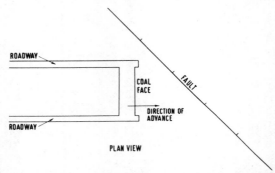

FIG. 4. Sketch of plan view of a face advancing towards an unknown fault.

has been encountered, it is necessary to decide whether to turn the face and advance parallel to the fault, whether to drive a new face on the other side of the fault and transfer the equipment, or whether to abandon the face altogether.

If the fault were to occur close to the end of the planned life of the face, the face would probably be abandoned. Work would begin on a new face which would probably be almost ready to start its life. There would be scarcely any break in production and the cost of the fault would be small. On the other hand, if the fault were encountered early in the life of the face, production on that face would stop for perhaps three or four months while the new face was being set up. It would not normally be possible in the meantime to make up for all the lost production by increasing production on the other working faces. There would be a net loss of production which would be, perhaps, 5% of annual output. If the colliery normally produced 1 million tonnes of coal per year to sell at £20 per tonne to break even, the annual costs would be £20 million and the cost of the face-stopping fault would therefore be £1 million.

Since the location of all faults with throw greater than 2 m is not normally known in advance, there is a risk associated with each face which is of the order of £1 million early in its life, and which diminishes with its progress. In order to insure against this risk, it is necessary to develop spare capacity, i.e. production faces which are either equipped but not working, or are working at less than their full potential. Manpower from an abandoned face can be transferred to increase production on these spare faces. However, the costs of carrying this form of insurance are very high. In some geological environments the risk of encountering geological disturbances is so high

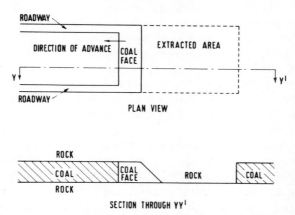

FIG. 5. Sketch to show the main features of a longwall retreat coal face.

that the cost of insuring against it can be too much. In other words, there are some geologies in which longwall mining is uneconomic. Obviously one wishes to avoid trying to mine coal in this way in such places.

If the risk of encountering face-stopping faults can be reduced, the cost of insuring against them can also be reduced. One way in which this problem is tackled in many modern mines is by retreat mining; the parallel roadways are driven out from the main roads first, and the coal is then extracted by retreating the face between the roadways as shown in Fig. 5. The risk of the face encountering unexpected geological hazards is reduced, for only after the geology has been 'proved' will the face be set up. Only the development will be abortive, not the whole face.

In its exploratory function a roadway is in principle no different from a lateral borehole and its ability to provide information about the structure around it is limited. Retreat faces are not therefore insured against encountering unknown faults.

In exactly the same way that seismic surveying is used in the oil industry to delineate geological structure, it can be used in the coal industry to the same effect. By locating major structures, a seismic survey can indicate how the faces should be arranged to minimise the risk of being stopped by unexpected geological difficulties. It can also reduce the risk of abortive development and can even reveal whether the intensity of faulting is too great to permit economic mining in the area, and thus save unprofitable investment.

So far this discussion of the effect of geological disturbance on mining costs has only been related to the year-by-year profitability of the mine, which also depends on the current price of coal. But equally important, and much more so in the long term, is the year-by-year reserves position of the mine. If the fault pattern is not well known in advance, then the initial layout of the mine may not be conducive to the easiest mining. The geological disturbances will be discovered in the course of mining, faces will be abandoned, new ones will be started up in different directions, and a best fit of mine layout to the geology may eventually be found. However, the extraction may be perhaps no greater than 30 % in any worked seam. If, on the other hand, the main geological features are known in advance, a better mine layout can be designed at the outset and a better fit of the rectangular panels within the fault pattern can be made. Since this will increase the percentage extraction in the seam (to perhaps 50 % in the same geology), the total output of the mine will be increased and the return on investment will also thereby be increased. *Furthermore, this course will also reduce the overall level of major capital investment in a nation's coal industry by deferring the need to sink new mines in order to maintain output.* Since the initial capital investment for a new mine is of the order of £80 per annual output tonne, i.e. £80 million investment for a 1 million tonne per year mine, the sums of money involved are enormous.

In summary, perhaps the biggest contributions that seismic reflection can make to the coal industry are,

(1) to permit a better fit of mine layouts to the geological environment in order to increase the percentage extraction of coal, and
(2) to reduce the risk of faces encountering unexpected faulting and hence to permit a reduction in the cost of spare capacity.

These contributions are likely to be cost-effective in appropriate geologies for all underground mining methods, especially those which are capital-intensive and cannot respond rapidly to unexpected geological disturbances. For a more comprehensive discussion of the effects of geological disturbance on mining costs see Clarke.[1]

The following is an outline of some of the ways in which the National

Coal Board has been applying seismic reflection to reduce the costs of extraction. It includes in particular an analysis of some of the problems which emerge when those field geometries normally used for oil and gas exploration are scaled down in order to examine the geological structure of importance to coalmining.

GEOMETRICAL CONSIDERATIONS

The depth at which coal can be extracted by longwall mining is limited by the induced stresses and by the temperature. In general, the deeper the coal is, the more difficult it becomes to maintain access roadways and to keep the roof and floor apart at the face. At the same time temperatures are higher and it can become too hot for men to work. In the UK the depth beyond which it is normally considered too difficult to mine coal is currently 4000 ft or 1300 m. This limit may not stand for all time. New methods of mining may be developed in which the induced stress conditions at present experienced at 1300 m would appear only at greater depths. If the coal at those depths were valuable enough, it would also be possible to install the necessary air conditioning to reduce the temperature sufficiently to permit men to mine it. However, at present, all the coal accessible to modern mining methods lies at depths less than 1500 m, and this fact places unavoidable constraints on the geometry of the seismic reflection system.

In conventional two-dimensional profiling one normally tries to design the length of the receiver spread to obtain the maximum subsurface coverage, both to reduce the cost of exploration and to obtain the best estimate of velocities to the target horizon. If the spread length is too short compared with the target depth, there are two adverse effects. First, the length of subsurface line covered by the spread will be correspondingly short and the number of records required to cover a given length of line will be too high. This will result in acquisition and processing costs which are too high. Secondly, the relative time delay, or moveout, of reflected arrivals across the spread will be too small and only a poor estimate of the average velocity to the target will be obtainable from the data. A reduction in costs and a better estimate of the velocities can be obtained by increasing the spread length. However, if the spread length is too long compared with the target depth two other adverse consequences usually arise. First, in a sedimentary sequence the velocity of sound has a tendency to increase with depth and above any target horizon there are usually layers of lower velocity. At a certain critical distance the refracted arrival from the base of a lower velocity layer will arrive at the same time as the reflection from the target, and obscure it. Secondly, even where the velocity structure does not produce this effect, the effect of normal moveout increases reflection arrival time approximately hyperbolically with distance (Dix[2]). When compensation for this effect is introduced in processing, there is an inevitable distortion, or 'stretch', of the reflected wavelet which degrades the resolution and continuity of events on the seismic section (Dunkin and Levin[3]). In processing, therefore, contributions from very large offset reflections are normally removed by 'muting' to avoid both these undesirable effects. Figure 6 illustrates this argument and shows the relation between subsurface coverage and maximum offset. It is clear that the depth of the target determines the spread length which is optimum in the sense that it maximises information content and therefore also minimises costs.

The maximum source–receiver distance which can normally be used to avoid losing reflection data through necessary muting in processing is approximately equal to the target depth. It follows that the maximum source–receiver distance can be no greater than about 1500 m for coal exploration. Any geophones which are further away from the shot than this will receive usable reflection only from below the zone of interest and are therefore wasted. Normally, of course, the coal target is much shallower than 1500 m and the spread length must be reduced accordingly.

The subsurface coverage is approximately equal to half the spread length for each shot. In common depth point shooting (Mayne[4,5]) each discrete subsurface point is covered a number of times, e.g. six-, twelve- or twenty-four-fold, to enable improvements in the signal–noise ratio to be made by stacking up reflections at varying angles of incidence. To obtain twelve-fold data at 1500 m with regular symmetric straddle shooting geometry the maximum shot-point interval cannot be greater than about 125 m. Quite often the target seams will be at less than half this depth, in which case the shot-point interval and geophone spread length must be reduced by the same factor, in order not to waste valuable recording channels. The only way in which such close shot-point spacing can be avoided is by reducing the fold of subsurface coverage; this can only be done if the signal–noise ratio is good enough on the majority of shots.

Quite simply then, because the zone of interest is shallow, relative to many oil and gas prospects, the shot-point interval is necessarily less and the *cost per km necessarily greater. Shallow data cost more to obtain at the same fold of coverage than deep data, all other things being equal.* This is a fact which is not willingly recognised and has been deliberately laboured here for that reason.

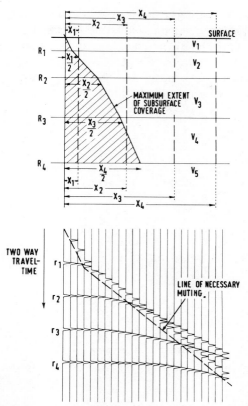

FIG. 6. Diagram to show the relationship between the maximum subsurface coverage and the depth of the target. The reflections on the record r_1, r_2, r_3 and r_4 correspond to the interfaces R_1, R_2, R_3 and R_4 respectively. No multiples are included. The line of muting defines the maximum shot–geophone distance for these reflectors as X_1, X_2, X_3 and X_4, respectively. The maximum extent of subsurface reflection coverage is shown by the shaded area.

The scale of a coal seismic section relative to a conventional oil seismic section is shown, very schematically, in Fig. 7. All the structure of interest to coalmining engineers would be seen within the first second of the seismic record, and in order to see more detail within this shallow part of the section it is necessary to increase the density of information there. It is not sufficient merely to sample the data more frequently in space and time, for this will reveal no further information if the data are already adequately sampled; it is also necessary to introduce more detail by generating higher frequencies.

Ziolkowski and Lerwill[6] discuss an approach to the production of higher resolution data by modifying the field technique. They conclude that there are some fairly simple modifications which should be made to the source and to the receiver. Explosive sources are favoured because they produce an unrivalled short high-energy pulse rich in high frequencies. The shot should be no bigger than necessary and placed below the weathering. The smaller

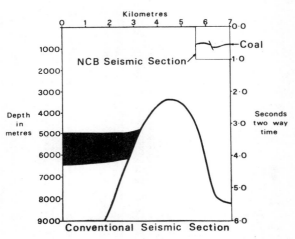

FIG. 7. Sketch to show the relative scale of a coal seismic section v. an oil section.

the shot, the more its spectrum is shifted towards the high frequencies, the smaller is its capability of exciting ground roll, and the greater is the potential resolution of the data. However, as the size of shot is reduced, the amplitude of the signal is also reduced and the ratio of signal to ambient noise is therefore also reduced. There is consequently a minimum charge size which can be used to good effect.

Ziolkowski and Lerwill[6] also conclude that the receiver spread should consist of a single phone for each station, rather than a pattern of phones. The principal reason for rejecting the pattern is that it will have a poorer high frequency response than a single phone due to two independent effects; the smearing effect of normal moveout across the pattern, and the destructive interference of high frequencies due to the unavoidable phase shifts introduced by differences in ground coupling.

In more conventional work the pattern has two purposes; to increase the output from the geophone station in order to overcome the cable noise, and also to discriminate against ground roll using the antenna effect. Since spread lengths for coal survey are relatively short, the cable noise may be reduced for no great increase in cost by shielding the cable connectors to reduce leakage. The ground roll is not treated so readily. The normal moveout of reflections across a given pattern is approximately inversely proportional to the depth of the reflector (assuming constant velocity). The attenuation of high frequencies due to this effect therefore becomes more severe when the target is shallow, but it is at these shallow depths that the high frequencies are needed the most. It is clear, therefore, that geophone patterns are unhelpful in the acquisition of shallow high resolution data and should therefore be avoided if possible. It follows that the attenuation of ground roll becomes a problem which is fundamental to shallow high resolution data acquisition on land and it will be considered in more detail in the next section.

It is worth noting that, since this technique relies on a single geophone per station, it depends critically on that geophone having very good coupling to the ground. In some areas it is easy to obtain a good response from almost any geophone. In other areas one is not so lucky and great care must be taken with the geophone plant. In such circumstances the design of the geophone case is almost as important as the fidelity of the phone. Where it is not possible to obtain good coupling it is sometimes necessary to use groups of phones in bunches. In these cases one is relying on statistics to produce a signal; the probability of obtaining some sort of signal is increased as the number of geophones per station is increased. In a good data area, however, bunches can be detrimental because the statistics can work in the opposite direction; any poorly planted phone in the bunch will increase the overall noise level disproportionately. Since it is impossible to analyse the contributions of individual phones to the summed output of a bunch, it is more difficult to identify and correct the noisy ones. There is, therefore, a tendency for groups to have a higher proportion of this non-correlated noise than single geophones.

It will be clear from this discussion that in order to focus the seismic system on the shallow structure, rather than on the deeper structures more typically of interest in oil and gas exploration, the geometry of the seismic system has to be scaled down accordingly. Unfortunately, it is inherent in this scaling down process that two kinds of noise which do not scale, namely, static errors and ground roll, have a magnified effect. The problems caused by their relative magnification are discussed in more detail in the next section.

STATIC ERRORS AND GROUND ROLL

Static Errors

Variations in the thickness and velocity of the surface layers, or weathering, cause variable delays in the arrival times of reflections from deeper layers. The weathering is normally of far lower velocity than the deeper rocks. Therefore small variations in its thickness produce time delays which could be produced deeper only by much larger structures. The distortions introduced into the reflected wavefronts as they pass through the weathering thus exaggerate the size of these variations relative to the deeper structure. Consequently deep structural features of far greater importance tend to become obscured. (There is not always a one-to-one correspondence between the seismic weathering which gives rise to this effect and the geologic weathering which can be identified by optical examination of the rocks.)

The magnitude of these variations and consequently the magnitude of the distortions they cause depend only on the survey area. They are quite independent of the geometry of the system and the depth of the target. Therefore the scaling down of the geometry required for shallow high resolution surveys has the effect of further magnifying the distorting effect of the weathering.

Normally, of course, corrections for these distortions are made by applying time shifts to the data which vary according to the shot and geophone positions. These shifts, or static corrections, attempt to simulate data which would have been obtained had the weathering not existed and had the shots and geophones all been placed on some convenient datum. In other words, the real earth is replaced by a model which has a simpler surface structure. However, in order to do this the differences between the model earth and the real one must be known rather accurately. If they are not, there will be significant errors in the applied time shifts. These are called static errors. Since the scaled down geometry magnifies the effect of the weathering, the static errors are magnified correspondingly. In order to reduce the absolute magnitude of these errors so that they may assume the same relative magnitude as they would in more conventional prospecting, it is essential to increase the accuracy of computation of static corrections.

What makes static errors particularly troublesome in coal exploration is that they tend to introduce apparent faults in the seismic sections. Figures 8 and 9 illustrate this point. They show the same piece of line with different multiplicities of subsurface coverage. Figure 8 contains all the data (a nominal twelve-fold coverage) and Fig. 9 contains only the data in which

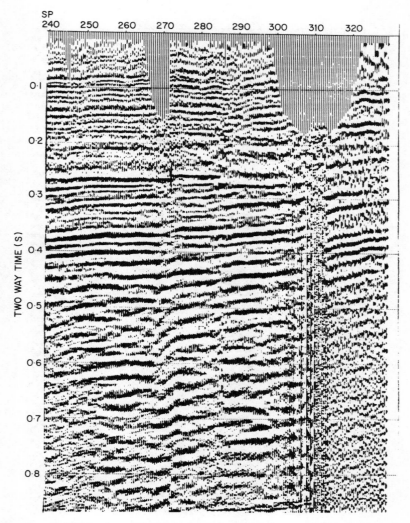

FIG. 8. 12-fold section with 100 ms between timing lines and 12 m between shot-points.

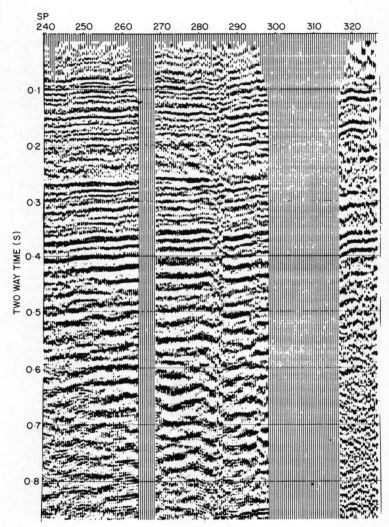

FIG. 9. Same section as Fig. 8, but distance-edited to exclude reflections obtained with an offset greater than 100 m.

the shot–receiver distance does not exceed 100 m (a nominal four-fold). The geophone station interval was 12 m. The inclusion of the longer offset shots in Fig. 8 allows the shot-point gaps to be filled in, but to some extent this has created more problems than it has solved. For example, in Fig. 8, underneath the gap at shot-point 270 (SP270) there is what could be interpreted as a steep normal fault, downthrowing to the left. In Fig. 9, there is no suggestion that anything of structural significance is present in the narrow shot-point gap. A close look at Fig. 9 will reveal, however, that the data to the right of the gap arrive earlier than the corresponding horizons to the left of the gap; there is a vertical time shift of about 10 ms or so between the two sides. This shift is caused by variations in the weathering which have not been taken into account in the calculation of static corrections. In Fig. 9 this is obvious. In Fig. 8, it is not so obvious *because the time shift is not entirely vertical.*

It is not vertical because the static errors have been smeared out across a number of traces. Any error in a shot or geophone static correction will contribute to the total static error of a number of traces in the stacked section. A shot-point static error, for example, will appear as the same fixed time shift on every trace from that shot. For regular shooting geometry and 24 channel recording, there would be 24 consecutive common depth point gathers each containing one trace from the shot with the static error. This error thus affects 24 consecutive traces in the stacked section. There are advantages and disadvantages in this.

One advantage is that random errors with zero mean will tend to cancel out with a high fold of coverage, because every common depth point gather will tend to contain traces with both positive and negative errors. The greater the fold of stack, the greater the probability that the static errors will average to zero. Of course, these errors will tend to smear events, broaden them and decrease the resolution of the data. However, the continuity of events tends to be stronger. This is clear from Fig. 8; the shallow data with low fold of stack are 'hairier' than the deep data with a higher fold of stack.

The big disadvantage when data with static errors are stacked is shown by the example of Figs. 8 and 9. Here a systematic error, caused by a specific near surface feature such as a step in the thickness of the weathered layer, is spread across a number of traces, giving the impression of deep structural disturbance. In this particular case the problem has been revealed by reducing the fold of stack and showing that there is a constant time shift of all horizons across the gap. There are many other more insidious examples which are much more difficult to detect.

Another obvious example of poor static corrections is in the larger shot-point gap of Fig. 8 where the continuity has been lost even though the gap has been closed by including long-offset shots. In this zone of poor continuity it is clear that the static corrections are bad, *but it is not clear whether some faulting is also present*; this is the problem.

In order to identify areas of coal which are of low risk as far as structural disturbance is concerned, some confidence limit must be put on the interpretation of the data. The data may allow all faults above a certain throw which intersect a seismic line to be identified. If there is any doubt about the number of inferred faults which may be caused by static errors, then the confidence in the fault pattern is reduced (especially since, from line to line, faults will have been correlated with static anomalies and static anomalies with each other). Furthermore the threshold of faulting which can be identified with confidence will have to be raised and, because this automatically defines the scale of faulting which cannot be excluded, the risk associated with all the areas surveyed is also increased.

In the vast majority of seismic data obtained by the National Coal Board to date (1979) the threshold of identifiable faulting is now limited, not by the frequency content of the data, but by the magnitude of the static errors. In other words, our experience has shown that it is often far easier to generate high frequency data with great potential for revealing structure in some detail, than it is to calculate the static corrections with sufficient accuracy to allow that potential to be realised. Consequently a great effort has been put into reducing static errors to a minimum. The main lines of our approach have been, briefly, as follows.

We begin by assuming that we have one or more layers of low velocity material overlying rocks of higher velocity. Within this group of lower velocity material, which we refer to collectively as the weathering there may, of course, be low velocity layers. We also assume that we have no prior information on the thickness of the weathering. Nevertheless, we wish to place our explosive charge below the weathering in order to be able to calculate static corrections in the manner illustrated in Fig. 10.

If the datum is chosen to be close to the shot and if the shot is below the weathering, the shot-point static correction, t_s, is simply,

$$t_s = -\frac{(E_s - E_d)}{V_2} \qquad (1)$$

where E_s is the elevation of the shot, E_d is the elevation of the datum, and V_2 is the velocity of the medium below the weathering. E_s and E_d can be obtained accurately from surveying and the datum is chosen such that

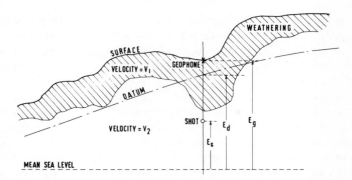

FIG. 10. Diagram to show relationship between elevation of shot, geophone and contoured datum. For accurate static corrections $|E_s - E_d|$ should be small and the shot should be below the weathering in fast material with velocity V_2.

$(E_s - E_d)$ is small. V_2 is estimated from refracted arrival times. Since $(E_s - E_d)$ is small and V_2 is large, t_s will be small. Any small percentage error in V_2 will make a correspondingly small percentage error in t_s.

The geophone static correction, t_g, is obtained from the shot-point static correction by including the measured travel-time of sound from the shot to the geophone, the 'uphole time', t_u,

$$t_g = t_s - t_u \qquad (2)$$

Data sampled at 1 ms intervals should probably be supported by static corrections accurate to 2 ms or less. If the error in t_u is of the order of 1 ms, and if the error in the estimate of V_2 is of the order of 10%, both $|t_s|$ and $|t_g|$ need to be less than about 20 ms.

It is crucial that the shot is not in the weathering (and steps should be taken contractually to ensure that it is not); if it is, then the denominator in eqn. (1) would be some unknown combination of V_1 (the average velocity of the weathering) and V_2 and would, therefore, be less than V_2; t_s would, on average, be larger, and errors in t_s larger. In practice to ensure that the shot is placed below the weathering one must determine the thickness of the weathering first, and then programme the drills to drill below it. Since the weathering can have rapid variations along the line of profile, it is necessary to find the depth of the weathering at every shot-point.

The most effective way to obtain this information is to have a refraction crew ahead of the main crew, working to obtain a continuous profile of the weathering. The velocities in the weathering and the velocity at the base of the weathering can be used to compute the weathering thickness (see Musgrave[7]). There will be an error associated with each estimate of the thickness and this should be taken into account in programming the drills. In the UK the velocity at the base of the weathering can vary from 2000 to 3000 ms, depending on the area.

It will be obvious from eqn. (1) that for t_s to be no greater than 20 ms the datum must not be more than 40 m above or below the shot, and it is not always easy to choose such a datum. Very often a plane, either horizontal or sloping, will not satisfy this condition, so it may be necessary to choose a contoured datum. If a contoured datum is chosen, it must be smooth enough not to introduce spurious structure into the seismic sections. One criterion for smoothness, which will normally be satisfactory for a contoured datum, is that the gradient of the datum must be essentially constant over distances less than the maximum shot–receiver distance. If this is the case the datum will not cause distortions in the horizons when normal moveout is applied to the common depth point gathers. However, contoured datums do present other difficulties especially in the migration of the data. This point will be discussed very briefly in the last section.

Even when a contoured datum is used, it is clear that there are some topographies in which it is not possible to satisfy simultaneously the smoothness criterion and the requirement that $|E_s - E_d|$ be less than 40 m, unless the maximum shot–receiver distance is reduced. There is thus a trade-off to be made between cost and velocity control on the one hand, and the accuracy of static corrections on the other. The balance point between these two is a matter of judgement, to be based on an assessment of the problem to be solved by the application of the seismic survey, and the value of the results. Halving the geophone spread length to permit a more rapidly varying datum to be used, for example, would reduce the magnitude of short-period static errors by perhaps 30%, but it would also reduce the velocity control and would result in approximately double the acquisition costs. In some situations the improved accuracy of static corrections would not be sufficient to justify these increased costs.

At first reading it might be argued that the contoured datum could be avoided altogether by choosing a single plane, below all the shots, and allowing the static corrections for points on the high ground to be larger. One could allow V_2 to change to match the near-surface geology and thus avoid distorting the time-structure. Provided the V_2 changes are made smoothly, no steps such as that at SP270 in Fig. 8 are introduced. There are two problems with this argument.

First, it assumes that variations in V_2 take place only laterally. However,

V_2 is only the velocity of the refractor at the base of the weathering. This refractor reveals nothing about the variation of velocity with depth of the deeper material between the base of the weathering and the datum. Failure to take both the lateral and this vertical variation of V_2 into account will result in distortion of the time structure. Since we do not know what the vertical variation in V_2 is, we cannot take it into account. Therefore we must use a datum as close to the shot as possible, within the constraints outlined above. V_2 is, of course, a smoothly varying local refraction velocity. (Incidentally, the step at SP270 in Fig. 8 was not introduced by a sudden change in V_2. It was caused by a sudden change in the thickness of the weathering not noticed in this early survey where shallow shot-holes were used.)

Secondly, large static corrections will result in a loss of shallow data, and these shallow high frequency reflections are often extremely useful in controlling the operation of automatic residual statics programs, provided the errors in the initial static corrections are small. The only way that such a loss of the important shallow data can be avoided when the static corrections are large is to introduce a large positive constant time shift to every trace, thus compensating for the larger corrections. But this operation incurs a penalty; it forces the trace-to-trace moveout of the reflections to be non-hyperbolic. Consequently correct stacking velocities become difficult to determine using Dix's formula[2] and the usual computer programs.

In short, the choice of datum is a non-trivial matter. Every datum has its own associated problems. One must choose a datum which will best enable the object of the survey to be achieved. In surveying for coal the primary object is to resolve structural discontinuities. Since their resolution is limited principally by the accuracy of the static corrections, our approach has been to design the field procedures and to choose the datum such that the trace-to-trace errors are minimised. But, even with such a devoted effort, there are always problems with the static corrections. These can be caused by difficult drilling conditions, unavoidable shot-point gaps, or indeed anything that prevents this ideal approach from being used.

In these circumstances a back-up procedure must also be used. Normally, this is an analysis of the first arrivals from a constant refractor corresponding to the base of the weathering. In multiple-fold common depth point data there is a redundancy of first arrival information which can be exploited to determine the static corrections with more confidence than would be possible with single-fold data.

Of course computer programs are also used to try to improve the static corrections automatically. However, the effectiveness of such programs is controlled not only by the signal–noise ratio (where the noise in this sense is not static errors) but also by the magnitude of the errors relative to the frequency content of the data. Thus a given program may just successfully 'correct' static errors of 16 ms in an area where the data are sampled at 4 ms intervals and where the dominant frequency of the data is 30 Hz. The same program will be unable to correct errors of the same magnitude in the same area where the data have been obtained with different techniques, sampled at 1 ms intervals and in which the dominant frequency is now 120 Hz. The scale of the errors has become too big for the bandwidth of the data. In order for such computer programs to be applied successfully to high resolution data, the static errors must already be small enough.

In summary, there are two reasons why very accurate field static corrections are essential in seismic reflection for coal. First, the presence of errors increases the threshold of detectable faulting; this means that areas of coal where the mining risk is low are identifiable with low confidence. Confidence can be increased if the threshold of detectable faulting is lowered, which in turn presupposes accurate static corrections. Secondly, the successful correction of static errors by computer programs depends on the magnitude of the errors relative to the frequency content of the data; since the frequencies of the shallow high resolution data are higher than those of conventional surveys, while the weathering problem is independent of the survey, a special effort must be made in the field to use a technique which will yield far more accurate static corrections.

Ground Roll

Ground roll is the name given to the surface waves generated by the sound source. The body wave reflections, which have travelled down to the horizons of interest and back up to the surface again, often arrive at the geophone at the same time as the ground roll. This ripples away from the shot much more slowly but travels the much shorter straight line distance along the surface. The geophone responds to the simultaneous arrival of the two waves in a way which, one hopes, is governed by a law of linear superposition. The geophone output, from the point of view of the reflection seismologist, therefore consists of reflection signals contaminated by surface wave noise.

The velocities and frequency content of the various possible surface modes are determined by the thicknesses and elastic properties of the material near to the surface. They are quite independent of the geometry of the seismic reflection system. Clearly the ground roll will be unaltered merely by scaling down the geometry of the seismic system. The same surface modes will be present.

Usually, of course, the ground roll energy is concentrated into a frequency band which is lower than the useful band of energy of the reflected signals. Therefore a certain amount of the ground roll can be filtered out simply by applying a low-cut filter, but because this can never have a perfectly sharp cut-off, it cannot attenuate the ground roll completely. In fact, if the ground roll is excessively large it can still drown the signal, even after the low-cut filter has been applied. This creates a problem in recording.

No recording system has infinite dynamic range. If the initial gain in set such that the amplifiers just fail to saturate, then the system will gain-range on the high-amplitude ground roll and may just fail to resolve the ripple of low-amplitude signal riding on it. If the recording filter is chosen to attenuate the ground roll more severely by rejecting higher frequencies as well, then it may also begin to cut into significant data. A better approach is to design filters with sharper cut-off characteristics. This has been adopted in some of the more modern recording systems.

Conventionally, ground roll is also attenuated in the field using a geophone pattern which discriminates between waves with different phase velocities. The pattern is designed to enhance waves with high phase velocities (the reflections) and to attenuate waves with low phase velocities (the ground roll). The geophone pattern thus acts like a filter (Savit et al.[8]) and can attenuate ground roll with wavelengths not greater than the length of the pattern.† Since the wavelengths present in the ground roll are independent of the scale of the field geometry, the patterns required to attenuate them are also independent of the scale of the field geometry.

At first sight, then, there is a fundamental difficulty in attenuating the ground roll with a scaled-down field geometry. However, if high frequencies are obtainable, there are several ways of attacking the problem.

First, the sound source must be made to generate as little ground roll as possible. If dynamite is used the amplitude of the ground roll it excites can be reduced if the shot is placed below the weathering rather than in it. This will also help in the calculation of accurate field static corrections.

---

† If all the geophones in the pattern have the same polarity, this is true. In order for the pattern to attenuate wavelengths longer than the length of the pattern, the polarities of the geophones must alternate. Furthermore, their excitations have to be very much greater than the output of the pattern so that the system is inefficient and often noisy (Tucker and Gazey[9]). For these reasons such 'superdirective' arrays are not used in the field.

Secondly, the amplitudes of the lower frequency longer wavelength modes of the ground roll can be reduced by reducing the charge size (Ziolkowski and Lerwill[6]). There are two reasons for this; the smaller charge radiates less energy, and this energy is shifted towards higher frequencies. Both these effects reduce the ability of the charge to pump energy into the lower frequency modes of the ground roll.

Thirdly, with the shift to higher frequencies, the sampling rate and the cut-off point of the high-cut filter can both be increased. There will normally be a corresponding absence of significant low frequency data which will permit a higher low-cut filter to be used in recording. This will increase the attenuation of low frequency ground roll.

The combined effect of these three factors is often enough to reduce the amplitude of the very low frequency ground roll, leaving the higher modes still to be tackled. But since these modes have shorter wavelengths, they may be attenuated in the field with a scaled down geophone pattern. In other words, the scaling down of charge size combined with the shift in recording filters to higher frequencies may attenuate the long wavelength ground roll so much that the significant remaining ground roll contains only such wavelengths as can be accommodated within the scaled down field geometry. Whether this happens to be the case or not will be a matter of luck, and will depend on the original distribution of energy between the various modes of ground roll.

Where scaled down geophone patterns are likely to be effective in attenuating ground roll for shallow high resolution surveys, the smearing effect of normal moveout will also scale down and will be important only at correspondingly higher frequencies.

Using this argument, one can almost convince oneself that ground roll is likely to be no more of a problem for shallow high resolution surveys than for more conventional surveys, but as the shift to higher resolution is pursued, the attenuating effect of the differences in geophone coupling within the pattern becomes more important, because it begins to take on significance at a fixed frequency which is independent of the scale of the pattern. In an experiment conducted with high fidelity phones, Ziolkowski and Lerwill[6] found this effect to be important at about 200 Hz. Of equal importance is the effect of the weathering. The static corrections for individual geophones often vary by several ms over a distance of 10 m. If a 10 m geophone pattern is used, a vertically travelling wave will arrive at a different time at each geophone within the pattern, the variation being of the order of ms. The pattern will have a consistent response at low frequencies and an inconsistent one at high frequencies above about

100 Hz. Thus if it is necessary to preserve reflected energy at frequencies much above 100 Hz, patterns must be rejected, even if those patterns are designed to attenuate ground roll at a much lower frequency, e.g. 30 Hz.

Once patterns are rejected ground roll does emerge, after all, as a problem of fundamental difficulty for high resolution seismic reflection. If both the reflections and ground roll can be recorded with sufficient accuracy within the dynamic range of the recording system, they may be separated in processing by the use of velocity filters. If the dynamic range of the recording system is too low to preserve both signals and noise then this is

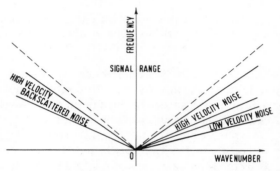

FIG. 11. A frequency–wavenumber plot to show the separation between high frequency low wavenumber reflections and lower frequency higher wavenumber noise.

not possible. At this point some compromise must be made; either to use geophone patterns and thus sacrifice some high frequency energy, or to use a higher low-cut recording filter and thus sacrifice some low frequency energy. Either way the bandwidth, and consequently the resolution, of the data must suffer.

A word should be said about the use of velocity filters. Figure 11 shows the general disposition of useful reflected wave signal and surface wave noise in a frequency–wavenumber plot. The velocity filter which needs to be applied would be defined by the dashed lines and should reject everything below these lines and pass everything above them (see Fail and Grau[10] and Treitel et al.[11]). Of course the field data are sampled at discrete intervals in space and time. There is thus a limit to the bandwidth in both frequency and wavenumber, and there are also problems with aliasing.

To prevent aliasing in time, a high-cut filter is applied before the signal is converted from analogue to digital, and this should be designed to remove

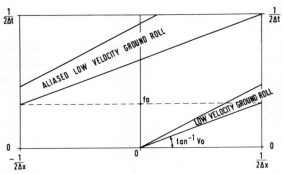

FIG. 12. A frequency–wavenumber plot for discrete data to show the wrap-round caused by high frequency low velocity noise aliasing into lower wavenumbers because the station spacing is too large.

all resolvable energy above the Nyquist frequency. With a single geophone per station, aliasing in space can only be controlled by the station spacing. If the station spacing is too large then the high frequency low velocity ground roll will alias back into the lower wavenumbers. This is illustrated in Fig. 12. The higher frequency ground roll 'wraps round' into the lower wavenumbers. The frequency, f_0, at which the wrap-round occurs is controlled only by the geophone spacing, ΔX, and the ground roll velocity, V;

$$f_0 = \frac{V}{2\Delta X} \qquad (3)$$

Aliasing of ground roll with velocity V will occur at frequency f_0 unless the

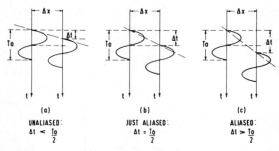

FIG. 13. The relationship between period, trace spacing and spatial aliasing. The moveout, Δt, between traces equals $\Delta X/V$, where V is the velocity of the event with period T_0, and ΔX is the trace spacing. The event is spatially aliased if $2\Delta t$ is greater than T_0.

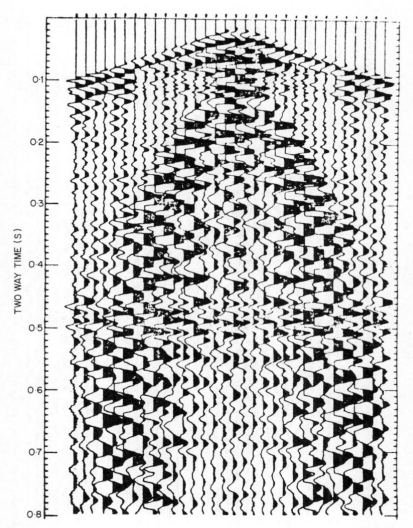

FIG. 14. A record showing reflections and ground roll. Trace spacing is 10 m.

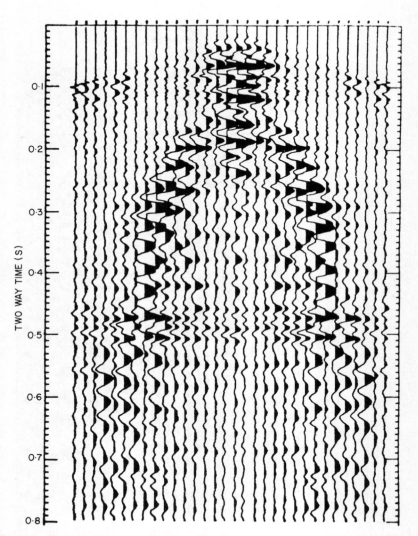

FIG. 15. Same record as Fig. 14 after velocity filtering in frequency–wavenumber space; aliased ground roll still remains.

geophone interval, ΔX, is reduced. If we define $T_0 = 1/f_0$ as the period at which aliasing occurs, we can see from eqn. (3) that

$$\frac{T_0}{2} = \frac{\Delta X}{V} \tag{4}$$

The relationship between period, trace spacing, velocity and aliasing is illustrated in Fig. 13.

An example of a record showing reflections and surface wave noise is illustrated in Fig. 14. This was obtained with 10 m geophone spacing and 1 ms sampling. The data were transformed by Kennett and Hughes[12] to frequency–wavenumber space and then filtered to pass events with apparent velocities greater than about 2 km/s, which is just less than the velocities of the first arrivals and consequently these are modified somewhat. The result is shown in Fig. 15.

There is good suppression of ground roll but a significant amount is still present on the filtered record, with a frequency of about 30 Hz. It is clear that to be able to reject this remaining ground roll with a velocity filter of the type mentioned above, a trace spacing no greater than about 5 m is required.

The point of this lengthy discussion is to come back to the field specification. The horizons of interest in Fig. 14 are at about 500 ms which corresponds to a depth of about 600 m. The maximum useful shot–geophone distance could therefore be as much as 500 m for example (see the discussion on geometrical considerations). If a straddle shot is used to maximise coverage and minimise costs, the spread length would be about 1000 m. But we have just seen that the geophone spacing should be no greater than about 5 m. Therefore, to record data from this area correctly, a recording system with about 200 channels is required.

This is perhaps an astonishing result, but it follows from the geometrical considerations and the rejection of geophone patterns. If very high frequency data are obtainable from depth, using perhaps buried hydrophones instead of geophones (Farr[13]), even more recording channels will be required. Until it becomes cost-effective to do this, compromises will have to be made in the field. These nearly always work out to be a trade-off between quality and cost.

INTERPRETATION AND PRESENTATION OF RESULTS

In all the foregoing, higher resolution has been described as a necessity but has not been defined. Farr[13] defines resolution in this context as 'the dimension of the thinnest single bed that will have a vertical reflection signal *equal in amplitude* to the signal from the single interface between the two semi-infinite mediums' (my italics). Arguing that the ability to detect a single interface will vary with the noise level and that if the single interface is detectable then at least a bed so defined will be detectable, he states that 'the minimum detectable bed thickness, as defined by this identical single amplitude criterion, is one-twelfth of the predominant seismic wavelength as determined from the bed velocity and predominant reflection frequency'. Of course, if the signal–noise ratio is greater, then thinner beds producing a lower amplitude reflection should be resolvable. How thin are the thinnest resolvable beds?

Kennett and Hughes[12] have computed some theoretical seismograms, using a wave theoretical approach, to determine the minimum bed thickness likely to give a detectable signal in the absence of noise. Without finding a non-trivial minimum, their results indicate that a bed only one-hundredth of a wavelength thick can produce a reflection, although its amplitude is small. This suggests that thin beds are detectable and do contribute to the seismogram. In a ray-theoretical approach, Rüter and Schepers[14] come to the same conclusion. The problem is that unless the elastic properties of the bed are known, and unless the reflection from the thin bed can be isolated, and unless some analysis of the amplitude is then performed, it will not be possible to determine from the seismogram the thickness of the bed or, indeed, whether it is the right bed.

Coal Measures, by their nature, usually contain cycles of thin beds, the coal seams being amongst the thinnest. Within the sequence of beds there are usually frequent extreme changes of velocity and density. The cyclic nature of the measures combined with the rapid changes of acoustic impedance give rise to a seismogram which contains many primary and multiple reflections. Some of the multiples are very short period interbed multiples and some are longer period. Kennett and Hughes[12] have shown that P-wave to S-wave and S-wave to P-wave conversions at the boundaries between the media are very important in the structure of the interbed multiples in these rapidly varying layered sequences.

The complexity of the reflection seismogram from Coal Measures has been investigated by, among others, Dresen,[15] Rüter and Schepers,[14] and Kennett and Hughes.[12] Beginning with a simple model, such as a single seam, and then adding seams to increase the complexity, they all reach the conclusion that after the first two or three seams the seismogram has become so complex that any alteration in the properties of any one seam, for example its thickness, cannot be deduced from the consequential alterations in the seismogram. But this is simply to say that the seismogram has low frequencies, whereas a higher frequency seismogram would perhaps allow small perturbations in, for example, the thickness of a particular seam to be resolved. Perhaps in time this may be possible. However, Rüter and Schepers[14] show that the cyclically layered sequence acts like a low-pass filter. The first five or six seams are such a powerful filter that detailed information from deeper seams is virtually unobtainable; if the seam of interest does not happen to be within the first five or six, the probability of obtaining high frequency reflections from it are likely to be poor. Of course the seams of economic interest are not always the shallowest ones in the sequence.

The importance of all this for the interpretation of seismic sections from Coal Measures is clear; it is very difficult to say in detail what the character of a seismogram reveals about the quality of any individual seam, but if that character remains constant such that correlations along the section are possible, then there must be continuity of some kind within the sequence. Figure 16 shows a seismic section obtained on Coal Measures with a cored borehole section plotted alongside. The borehole log has been plotted in two-way travel-time, using velocity data from the borehole. The Coal Measures sequence begins at about 0·41 s, and it is clear that it dips gently to the right. It is tempting to make correlations with individual seams (especially since coal is black and the reflections are black) but such correlations may well be fortuitous in view of the complex inter-relationships between primary and multiple reflections within the sequence. Therefore, we would not expect to be able to infer that the seam of interest was thickening or thinning, or splitting or changing in ash content, although with the benefit of hindsight and some borehole information we might try to convince ourselves afterwards that a detectable change in character was attributable to one of these things (and not attributable to any of the changes in any of the other seams, or even to changes between seams). In other words, there is as yet not enough known about coal seismic sections to permit variations in seam properties to be predicted from changes in character. Boreholes are still required to determine whether the coal is present, whether its quality is still good, better, or worse, whether the strength of the roof and floor has changed, and so on. So far, seismic reflection can still only be used to determine structure, as was emphasised in the introduction.

But determination of structure can be very critical, as we saw. From Fig. 16 it can be seen that the Coal Measures dip to the right and that there is very little evidence of faulting, except possibly below 0·7 s, where the

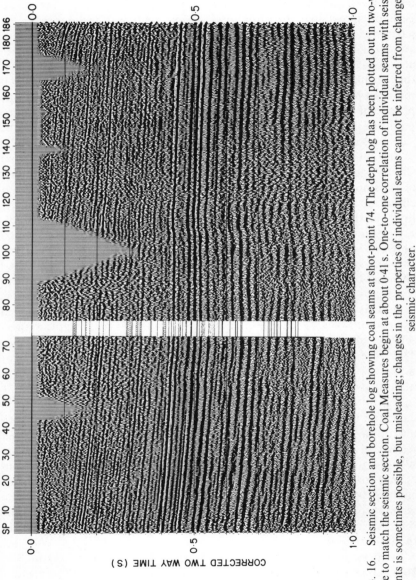

FIG. 16. Seismic section and borehole log showing coal seams at shot-point 74. The depth log has been plotted out in two-way time to match the seismic section. Coal Measures begin at about 0·41 s. One-to-one correlation of individual seams with seismic events is sometimes possible, but misleading; changes in the properties of individual seams cannot be inferred from changes in seismic character.

signal–noise ratio is less good. This information can be put onto a map to help the mining engineer. If velocities are known it is also possible to convert the contoured time information into contoured depths, and it is here that a significant difference arises between interpretation for oil and gas on the one hand and coal on the other.

In some oil and gas prospects, where the structural gradients are small, extremely accurate velocity information is often required to determine the positions of anticlines, and possible reservoirs. Without this accurate velocity information whole oilfields can be lost through incorrect time-to-depth conversion. In such structures the oil is obtained only in anticlines, where it has accumulated through a long process of permeation under pressure, because it is fluid. Coal, being solid, will not move in this way. Minor changes in the gradient do not affect its continuity, and mining systems are also able to accommodate these minor changes without changes in costs. Therefore, because the economic consequences of small changes in gradient are far smaller for coalmining than they are for oil prospecting, the need for accurate velocities is not quite so pressing.

However, as we noted earlier, operational mining costs can be dramatically affected by faulting, even faults with throws as little as 2 m can have a profound effect. The primary use to which seismic reflection is put in the exploration for coal in the UK is therefore to determine structure in as much detail as the data will permit. The detail is limited, as we saw, by the accuracy of the static corrections, as well as by the bandwidth. Within these limits each seismic line is able to resolve faults above a certain throw; this piece of information, determined from the data quality, allows a geological judgement about the mining risk to be deduced.

In a particular area there may be a known main trend of faulting, e.g. NE–SW. The rate at which faults grow from nothing to maximum throw may vary within the area (and very little may be known about this statistically; there may only be a few isolated examples), but there may be some likely maximum rate of growth, e.g. 1 m in 40 m. In other words, the majority of faults in the area may trend NE–SW and grow at a rate of not more than 1 m in 40 m. A 10 m fault is likely to have a total strike length of at least 800 m. Conversely, if two NW–SE lines are 400 m apart and no faults are seen to cross these two lines then the area in between probably contains no faults with throw greater than 10 m. Furthermore, to the NE and SW of these two lines there are areas parallel to the lines, extending up to 400 m on either side, which probably do not contain faults with throws greater than 10 m.

Using such geological assumptions and the seismic data it is possible to present on a map not only the interpreted structure, *but also areas on the map where faulting up to a certain maximum throw could exist and remain undetected by the grid of seismic lines.* Where there is an absence of lines or boreholes, or where the data quality is bad it is not possible to assess the mining risk more accurately than knowledge of the regional geology will allow. Whether it is worth finding out more about these areas depends on their possible effect on the mine plan. If the mine plan will be unaffected by knowledge about these areas then any exploration there will be pointless. If exploration is required, and the predominant fault strike direction and maximum rate of growth are known, there will be an optimum line spacing for a given data quality (where 'quality' is defined as the ability of the data to reveal small faults).

It was said in the section on static errors and ground roll, that the ability of the data to resolve small faults was usually limited by the accuracy of the static corrections, and that these could often be improved by the use of a contoured datum. A contoured datum always causes problems with the accurate positioning of horizons in depth. If the datum is converted to a plane before applying normal moveout and stacking by assuming a velocity structure between the datum and the plane, this is likely to make the velocity determinations difficult and result in a poor stack, especially if the assumed replacement velocity structure is incorrect. (It will be very difficult to obtain hard information on the probable replacement velocity structure, so this is not likely to be correct.) If the datum is converted to a plane after applying normal moveout and stacking, by assuming a slowly varying replacement velocity structure and applying the appropriate time shift to each trace, the horizons will be bent into new positions in time. Their migration to depth will be affected by these additional time shifts whose accuracy will depend on the accuracy of the assumed velocity structure. However, an important point is that even if the assumed replacement velocity structure is wrong, it will not introduce small faults where none existed, nor will it remove any which did exist, because the bending of the horizons will be smooth. Therefore, if minor faulting can be detected more easily by the use of a contoured datum, it should be used.

One should not retain the impression that accurate velocities are not important for coal seismics. They will affect the final interpreted position of major and minor structures, and consequently the more accurately they are known, the more accurate is the interpretation. It is exactly the same as for oil and gas. Nevertheless, since the economics of coalmining and oil exploration are affected in different ways by geological structure, one should place most emphasis on those aspects of the seismic system which will have the greatest benefits to the industry being served. It is the aspects which most affect the coalmining industry that have been emphasised here.

ACKNOWLEDGEMENTS

Such understanding as I have of the threads of the arguments presented here I owe to the help of many kind friends; the flaws are all my own work. I am indebted to Mr A. M. Clarke, Chief Geologist of the National Coal Board, who, probably more than anyone else, saw the need nearly two decades ago to bring the technology of geophysics into the coal industry to solve mining problems, and has campaigned for its use and development ever since. Dr A. A. Fitch in a very prescient letter gave me an early appreciation of the problems, particularly the difficulty of obtaining good static corrections, even before I began to work for the National Coal Board.

I would also like to thank Mr J. M. Slater, Mr C. P. Eaton, Mr M. J. Allen, Mr T. M. Jowitt, Mr R. Slack, Mr R. H. Hoare, Dr G. B. Barnsley and Mr R. E. Elliott of the National Coal Board; Mr W. E. Lerwill, Mr D. E. March, Mr D. F. Christie, Mr J. M. Richards, Mr A. Woolmer, Mr J. Elder and Mr A. Ward of Seismograph Service (England) Ltd; Dr J. B. Farr, Mr D. Brown, Mr Q. Williams and Mr D. Skerl of Western Geophysical Co. of America; Mr A. Tilsley of Geophysical Data Analysis; Mr D. Popovic of Geomatic; Dr B. L. Kennett, Ms V. Hughes and Mr J. C. Dreyfus of the Department of Geodesy & Geophysics, University of Cambridge; Mr R. Griffiths of Seismic Geocode Ltd; Dr C. W. Frasier and Dr T. E. Landers of Lincoln Laboratory, Massachusetts Institute of Technology.

I thank the National Coal Board for permission to publish this contribution; however, the views expressed in it are my own and are not necessarily those of the Board.

REFERENCES

1. CLARKE, A. M., Why modern exploration has little to do with geology and much more to do with mining, *Colliery Guardian Annual Review*, Aug., 1976.
2. DIX, C. H., Seismic velocities from surface measurements, *Geophysics*, **20**, No. 1, p. 68, 1955.
3. DUNKIN, J. W. and LEVIN, F. K., Effect of normal moveout on a seismic pulse, *Geophysics*, **38**, No. 4, p. 635, 1973.

4. MAYNE, W. H., Seismic Surveying: US Patent 2 732 906 (application 1950) Abstract; *Geophysics*, **21**, p. 856, 1956.
5. MAYNE, W. H., Common reflection point horizontal data stacking techniques, *Geophysics*, **27**, No. 6, p. 927, 1962.
6. ZIOLKOWSKI, A. and LERWILL, W. E., A simple approach to high resolution seismic profiling for coal, *Geophys. Prospecting* (in press).
7. MUSGRAVE, A. B., *Seismic Refraction Prospecting*, Section 4, Society of Exploration Geophysicists, Box 3098, Tulsa, Okla., USA, 1967.
8. SAVIT, C. H., BRUSTAD, J. T. and SIDER, J., The moveout filter, *Geophysics*, **23**, No. 1, p. 1, 1958.
9. TUCKER, D. G. and GAZEY, B. K., *Applied underwater acoustics*, p. 199, Pergamon, London, 1966.
10. FAIL, J. P. and GRAU, G., Les filtres en eventail, *Geophys. Prospecting*, **11**, p. 131, 1963.
11. TREITEL, S., SHANKS, J. L. and FRASIER, C. W., Some aspects of fan filtering, *Geophysics*, **32**, No. 5, p. 789, 1967.
12. KENNETT, B. and HUGHES, V., Private communication, Dept of Geodesy and Geophysics, Cambridge University, Cambridge, UK, 1978.
13. FARR, J. B., High resolution seismic methods to improve stratigraphic exploration, *Oil Gas J.*, Nov. 1977.
14. RÜTER, H. and SCHEPERS, R., Investigation of the seismic response of cyclically layered carboniferous rock by means of synthetic seismograms, *Geophys. Prospecting*, **26**, p. 29, 1978.
15. DRESEN, L., Modellseismische Untersuchen zum Reflexions verhalten eines einfachen zyklisch geschichteten Steinkohlengebirges, *Glückauf-Forschungshefte*, Dec., 1976.

INDEX

Absorption, 48
Acceleration, 207 et seq.
Acoustic impedance, 127
Acoustic velocity log, 93, 100
Acoustilog, 95
Adaptive mode, 256 et seq.
Admittance, 126
Air guns, 97, 191, 196, 253
Alias, 295 et seq.
Ambac Industries, 216
Anisotropy, 52 et seq.
Aquaflex, 156–7
Aqua-pulse, 229
Aquaseis, 156
Array, 49, 73 et seq., 98, 196 et seq., 293
Astronomic north, 217
Atomic nucleus, 221 et seq.
Autocorrelation, 138, 247 et seq.
Automatic gain control, 45
Average velocity, 3, 17 et seq., 94, 111, 279

Ball bearing, 211
Baseplate, 116 et seq.
Beam
   position computer, 226
   slope, 211
Bell accelerometer, 216

Bias, 13 et seq.
Black powder, 144 et seq.
Bord and pillar mining, 273
Borehole fluid, 95
Bouguer corrections, 219, 228 et seq.
Bright spot, 253
Bubble pulse, 76, 151, 155 et seq.
Buoyancy compensating cell, 218

Calibration (of velocity logs), 111
Carbonate facies, 65
Cassios water gun, 181
Caving, 102
Cavitation, 163, 190
CDP velocity, 3 et seq.
Centrifugal outward effect, 216
Chalk, 260
Coal, 271 et seq.
   measures, 249, 299–301
Coefficient, 210
Coherence, 3, 9, 12, 20 et seq., 257
Common depth point (CDP), 2 et seq., 76, 280, 287
Compliance, 119, 126, 130
Compressional wave, 54
Constant velocity stack, 28
Continuous velocity log (CVL), 95
Convolution, 110, 241, 246
Correlation, 64 et seq., 240 et seq. 288, 301

Counting circuit, 223
Coupling, 283
Cross
    accelerometer, 226
    correlation, 22, 134, 136, 250 et seq.
    coupling, 210, 213, 216
    gyro, 226
    horizontal axis, 226
    torque motor, 226
Cycle skipping, 28
Cyclic bedding, 249, 263–4
Cyclothem, 249

Damping, 121, 126 et seq., 209 et seq.
Datum correction, 45
Decca, 224
Decomposition, 251
Deconvolution, 59, 201, 250 et seq.
Decoupling, 199
Deterministic deconvolution, 201, 253 et seq.
Deviated well, 111
Differential shift, 103
Diffraction, 27 et seq., 55
Digital
    recording, 207
    tape, 100
Dinoseis, 163, 194
Dip-bar display, 28
Dipping reflector, 18, 54 et seq.
Distance weighted pattern, 69
Diurnal effect, 223
Downgoing waveform, 256 et seq.
Downward travelling event, 106 et seq.
Dynamite, 76, 115 et seq., 144, 146 et seq., 242

Electrical discharge sources, 181
Electronic picking, 106
Element weighted pattern, 69
Elevation correction, 218
Eötvös correction, 216 et seq., 225, 228 et seq.
Epoch, 244

Errors (of velocity), 41 et seq.
Evaporite series, 102

Facies change, 30
Fast Fourier transform, 69
Fathometer, 207, 219, 226, 228 et seq.
Feathering, 42 et seq.
Feedback, 213
Filtering, 209, 216, 225, 241 et seq.
First break, 105, 240
First trough, 105 et seq.
Fish kill, 144, 149
Fixed radio patterns, 224
Flexichoc, 164
Flexotir, 154
Fluid content, 93, 242
Fourier transform, 69
Free air correction, 228
Freely swinging gravity meter, 206

Gal, 206
Gamma, 206, 221
Gas gun, 157, 160
Geoid, 218
Ghosts, 242, 245, 247
Goaf, 273–4
Graf, 215
Gravitational torque, 209
Gravity meter, 206
Ground roll, 76, 282 et seq., 292 et seq.
Group velocity, 105
Gyroscope, 206, 213, 225, 228 et seq.
Gyro-stabilised platform, 206

Heave, 207
Homomorphic deconvolution, 255
Horizontal accelerometer, 206, 213
Hydrosein, 163, 194
Hyperbolic
    approximation, 10
    truncation, 10

Impulse generators, 116
Inertial navigation, 225
Instrument filter, 254

International Geophysical Reference Field (IGRF), 221, 228 et seq.
Interval velocity, 17 et seq., 94, 240
Inversion, 250 et seq.

Jitter, 46

La Coste and Romberg, 209, 213 et seq.
Lanes, 224
Larmor precession, 222
Lateral velocity variation, 113
Latitude correction, 218, 228 et seq.
Levinson recursion, 252
Limestone, 137, 266
 -shale alternation, 249
Linear
 bulk shift, 103
 pattern, 69, 73
 prediction, 246 et seq.
Lithological change, 93, 240
Lithology, 64 et seq.
Logging tool, 100
Longwall mining, 273, 277
Lorac, 224
Loran C, 225 et seq.

Macro-anisotropy, 52
Magnetic
 air-gun, 181
 declination, 221
 field, 221 et seq.
 field strength, 221
 tape, 100
 units, 221
Magnetometer, 206, 209, 221, 223
Marine seismic source, 143 et seq.
Mass–capacitance (M–C) analogy, 120
Mass–inductance (M–L) analogy, 120
Maximum coherency stacking (MCS) velocity, 3 et seq.
Maximum entropy spectral analysis (MESA), 256 et seq.
Maxipulse, 152, 194

Mechanical impedance, 117, 127
Meter zero, 218
Mica water gun, 180
Micro-anisotropy, 52
Migration, 63, 66, 240
 velocity, 63
Milligal, 206
Mine lay-out, 278
Mini Sosie, 140
Minimum phase, 244, 250, 253 et seq.
Minisleeve, 185, 196
Move-out
 effect, 113, 240, 279
 velocity, 3
Mud travel path, 96
Multichannel predictive adaptive Burg algorithm, 258
Multiple reflection, 27, 53 et seq., 144, 242, 245, 299
Multiplexer, 44
Multipulse, 147
Muting, 280

Navigation, 233 et seq.
Niagaran group, 259
Nitro-ammonium nitrate, 147
Nitrocarbonitrate, 152
Noise, 41, 45 et seq., 245
Normal moveout, 11 et seq., 280, 282, 304
 velocity, 3
Normalisation, 45
Nyquist frequency, 296

Oersted, 206
Offset source survey, 111 et seq.
Opencast mining, 273
Overburden, 273
Overpressured zone, 66, 93

PAR air-gun, 169
Parallax, 36 et seq.
Pattern, 69, 73 et seq., 282, 293 et seq.
Peg leg multiple, 54, 245

Penta erythritol-tetranitrate (petn), 156
Period, 170, 215, 299
Phase-sensitive rectifier, 216
Phase velocity, 105
Photo cell, 211
Pinch-out, 258
Porosity determination, 93
Power spectrum, 40, 254
Precess, 213, 222
Precision errors, 62
Prediction error, 246
Predictive distance, 251 et seq.
Predictive error filter, 251 et seq.
Primary reflections, 53 et seq., 110 et seq., 299
Programmed gain control, 45
Propagation velocity, 41
Propane, 157, 161, 163, 185, 196
Proton precession magnetometer, 206, 221
Pull-up, 64
Pulse broadening, 105

Quasi anisotropy, 52

Rammer, 140
Range–range, 224
Raydist, 224
Rayleigh–Willis formula, 173, 187
Reaction mass, 117 et seq.
Reefs, 240, 259 et seq.
Reflection coefficients, 107, 241 et seq.
Reflectivity, 241 et seq.
Retreat mining, 277
Reverberant system, 113, 242, 245, 247, 251
Ricker wavelet, 246 et seq.
r.m.s. velocity, 3, 7 et seq.
Rock velocity, 64

Sal log, 229
Salt, 30, 102
Sands, 240, 261
Sand–shale series, 65, 102, 249

Satellite navigation, 224 et seq.
Seismojet, 174
Seisprobe, 157
Semblance coefficient, 23
Sentinel far-field recording system, 200
Shale, 30, 65
Shear wave, 55
Shifting stack, 51, 53
Ship motion, 42
Shoran, 229
Side swipe, 28
Sidereal day, 217
Signature (of seismic source), 98, 144, 152, 156 et seq., 253 et seq.
Simplon water gun, 175
Skewness, 46
Sonar Doppler, 224 et seq.
Sonic log, 95
Sparker, 181
Spectrum whitening, 251
Spherical divergence, 107
Spike, 143 et seq., 202
Spiking deconvolution, 250 et seq.
Spring tension, 210, 226
  control unit, 226
Stable platform gravity meter, 206, 213
Stacking velocity, 3, 20 et seq.
Statics, 30 et seq., 283, 284 et seq.
Stationarity, 256
Still reading, 218
Straddle spread, 280
Straight line La Coste and Romberg Meter, 213 et seq.
Stratigraphic trap, 240
Stratigraphy, 64, 253, 259
Strip chart, 226
Structural deconvolution, 255 et seq.
Structural traps, 240
Subjective errors, 60
Superseis, 152
Surface ghost reflection, 106
Swept frequency, 116 et seq.
Synthetic seismogram, 93, 245

Tail (of reflection), 110

INDEX

Tapering function, 258, 268
Terrain correction, 219
Tesla, 206
Thin layers, 48, 240, 242, 258, 261, 299
Third axis gyro, 225
Thumper, 140
Time–depth function, 107, 303
Toran, 224
Torque, 213 et seq.
Transductance, 125
$T \Delta T$ method, 2
T^2-X^2 method, 2

Unconformity, 64
Upgoing waveform, 256 et seq.
Upward reflected waves, 107 et seq.

Vaporchoc, 165, 194, 253
Varian magnetometer, 223, 226, 229
Velocimeter, 229
Velocity
  analysis, 3, 20, 22 et seq.
  detonation, of, 146
  distribution, 93
  filters, 295
  heterogeneity factor, 10, 14 et seq.
  profile, 28, 34, 37 et seq.
  spectrum, 21, 23, 27 et seq.

Vertical seismic profile (VSP), 106 et seq.
Vibrators, 116 et seq.
Vibroseis, 98, 116, 134, 136, 181, 242, 253 et seq.

Wall-clamped well geophone, 97
Water bottom multiples, 76
Water depth correction, 218 et seq., 228 et seq.
Wavelet, 47 et seq., 135, 144, 242 et seq., 244 et seq., 280
  processing, 254 et seq.
Weathered layer, 242, 282, 284, 287 et seq.
Weight-drop, 134, 137, 139, 140
Weighted average velocity, 16
Well geophone, 97, 99
White noise, 268
Whitening, 44
Wiener–Hopf filter, 247, 258
Wiener–Levinson algorithm, 268
Wiener shaping filter, 201, 247 et seq.
Wrap-round, 296

Zero-length spring, 209 et seq.
Zero phase, 240, 254, 262, 264